I0474886

High School Algebra

Dedicated to my team

Copyright © 2011 by Chandramouli Mahadevan,
on behalf of Astrarka
All rights reserved.

No part of this book may be reproduced, stored, or transmitted by any means, whether auditory, graphic, mechanical, or electronic, without written permission of both publisher and author, except in the case of brief excerpts used in critical articles and reviews. Unauthorized reproduction of any part of this work is illegal and is punishable by law.

ISBN-13: 978-1463715458 (CreateSpace-Assigned)
ISBN-10: 1463715455
BISAC: Mathematics / Algebra / Intermediate

Foreword

We wanted to start off the discussion with a short biography of Algebra. We believe that this will help to set the stage to understand the topic more intimately, as opposed to treating this as yet another endurance test with lots of problems to solve and several more symbols and shapes to deal with. We have relied on Wikipedia for creating this sketch[1].

In ancient Egypt and Babylon, people originally learned to solve linear ($ax = b$) and quadratic ($ax^2 + bx = c$) equations, as well as *indeterminate equations* such as $x^2 + y^2 = z^2$, whereby several unknowns are involved. The ancient Babylonians solved arbitrary quadratic equations by essentially the same procedures taught today. It was amazing that they could also handle a few indeterminate equations.

The Alexandrian mathematicians Hero of Alexandria and Diophantus continued the traditions of Egypt and Babylon, but Diophantus's book *Arithmetica* is on a much higher level and gives many surprising solutions to difficult indeterminate equations. So, it is likely that most of us treat the work of Diophantus as the source that dealt with indeterminate equations.

This ancient knowledge of solutions of equations in turn found a home early in the Islamic world, where it was known as the "science of restoration and balancing." The Arabic word for restoration, *al-jabru*, is the root of the word *Algebra*. In the 9th century, the Arab mathematician al-Khwarizmi wrote one of the first Arabic algebras, a systematic exposé of the basic theory of equations, with both examples and proofs.

By the end of the 9th century, the Egyptian mathematician Abu Kamil had stated and proved the basic laws and identities of algebra and solved such complicated problems as finding x, y, and z such that $x + y + z = 10$, $x^2 + y^2 = z^2$, and $xz = y^2$.

---

[1] *Refer to http://en.wikipedia.org/wiki/History_of_algebra*

Ancient civilizations wrote out algebraic expressions using only occasional abbreviations, but by medieval times Islamic mathematicians were able to talk about arbitrarily high powers of the unknown x, and work out the basic algebra of polynomials (without yet using modern symbolism). This included the ability to multiply, divide, and find square roots of polynomials as well as knowledge of the binomial theorem. The Persian mathematician, astronomer, and poet Omar Khayyam showed how to express roots of cubic equations by line segments obtained by intersecting conic sections, but he could not find a formula for the roots. A Latin translation of Al-Khwarizmi's *Algebra* appeared in the 12th century. In the early 13th century, the great Italian mathematician Leonardo Fibonacci achieved a close approximation to the solution of the cubic equation $x^3 + 2x^2 + cx = d$. Because Fibonacci had traveled in Islamic lands, he probably used an Arabic method of successive approximations.

An important development in Algebra in the 16th century was the introduction of symbols for the unknown and for algebraic powers and operations. As a result of this development, Book III of *La géometrie* (1637), written by the French philosopher and mathematician René Descartes, looks much like a modern algebra text. Descartes's most significant contribution to mathematics, however, was his discovery of analytic geometry, which reduces the solution of geometric problems to the solution of algebraic ones. His geometry text also contained the essentials of a course on the theory of equations, including his so-called *rule of signs* for counting the number of what Descartes called the "true" (positive) and "false" (negative) roots of an equation. Work continued through the 18th century on the theory of equations, but not until 1799 was the proof published, by the German mathematician Carl Friedrich Gauss, showing that every polynomial equation has at least one root in the complex plane.

The development of varies branches of Algebra continued through the 16th century through to the 19th through contributions from mathematicians in England, Germany, France and Italy.

The essence of the hardwork of our forefathers in Mathematics has helped us to formalize several concepts and tools that we has come to use. I remember an exercise in my school days of trying to explain induction without using any formulas or modern day tools and representation. This exercise taught me to value the work of the Mathematicians. Things that we take for granted have been drilled into our thought process - during our early years at school. Our contribution and our relationship to this field depends upon our ability to internalize these concepts and master the methods of formal problem solving.

Thus, while learning and understanding the concepts and definitions are important, one can build up expertise in Mathematics only by solving problems. Mathematics requires a combination of two skills —comprehension of concepts and related nuances and the ability to recall the appropriate tool, technique, formula or method in madness to solve a problem at hand.

We sincerely hope that the student is able to get a good grasp of the subject and the techniques after working with the content of this book. If the experience of going through this work is joyful for the student and works as a tool for building his / her understanding, we would be satisfied that we have met the primary objective of this effort.

Chandramouli Mahadevan
Astrarka.
Bangalore, India.

Preface

This book is an integral part of a 3 volume series on High School Algebra.

The book "High School Algebra" covers the concepts involved in the various topics of this subject. A few selected problems are solved after each chapter, to aid the understanding of the student. The book finishes with a collection of problems that the student must practice on, to gain expertise.

"Problems in High School Algebra" is a comprehensive solution set to the battery of over 1500 problems in all topics covers in the first volume. The student is expected to make an honest attempt to solve the problems before looking at the suggested solutions. These solutions are systematic and comprehensive. No intermediate steps are skipped; which ensures that the overall flow of the problem solving process starting with the initial conditions to the final solution is maintained.

Finally, there are students with an instiable urge for problem solving. "Challenges in High School Algebra" is intended to address this requirement. Over 250 odd gems on the topic have been covered.

The best way to use this book is for the student to attempt each problem on his/her own. In doing so, the depth of understanding in the subject improves. Mathematics is not a spectator sport. It requires patience, perseverance and practice. The level of expertise in the subject in some sense is directly proprotional to the number of problems solved by the student. The term "solved" is used to imply accuracy of thought, stringing together intermediate steps and accuracy of the final result. In a way, this term refers to the quality of the means and the quality of the end goal for each problem.

This work is a comprehensive self study guide for the students who desire to improve their understanding, appearing for Mathematics related competitive examinations and tests. These works are based on the gold standard on the topic by Prof Hall and Prof Knight. They published the book in late 1800s. This forms the central reference in several schools and colleges across the globe.

I believe that Astrarka has been blessed to have had the opportunity to work with some of the best and brightest Any work of this magnitude is always a product of teamwork. R Balasubramanian, Shilpa Jaikumar and Venkatratnam Pandit have contributed a great deal to this effort. A big thanks goes to the family members of our team. They have been a great source of inspiration during this entire effort. They have made a personal sacrifice to ensure that Astrarka succeeds. Without the unflinching commitment and single minded dedication of my team and the members of their family, this book would have been an exercise in futility.

Chandramouli Mahadevan

Table of Contents

1 Good Habits

There are five fundamental principles, or say good habits that we would like to emphasize before we commence our discussion on High School Algebra.

1. Neatness is conducive to accuracy. Refrain from the temptation to write down something quickly and then scratch the same to make the necessary corrections.

2. One of the weaknesses we find in students while solving word problems is the usage of $=$ sign. This sign has a specific meaning in the world of mathematics. It cannot be used as a way to begin every new line or step in the problem solving process. Use appropriate mathematical signs and symbols. Never use them to mean something vague. $=$ Sign is never good space filler.

3. Spend a second or two to explain how you arrived at a certain step. Several books and references use a statement, such as ``it follows from the above statement". We have oftentimes wondered how the expression or equation below follows from the one above. A good explanation is an excellent demonstration of your understanding of the underlying principles.

4. When you are faced with several conclusions during a problem solving process, it is a good idea to number the statements or equations. In subsequent steps, you can refer to these conclusions by using the label or the assigned equation number.

5. The easiest of problems attracts the silliest of mistakes. If the problem is easy, motivate yourself to get it right. Do not let overconfidence or carelessness take control of the situation.

2 Introduction

To say that Algebra is useful, therefore, we must learn it, is an understatement. This book focuses on problem solving strategies. We have organized the material into problems, the solution of each problem immediately after the statement. Familiarity with middle school arithmetics and elementary algebra is assumed.

This book must not be read like a work of fiction. Instead, the student is advised to spend quality time in ensuring conceptual understanding. Mathematics requires three skills. Let us recall these.

Comprehension: At the core of Mathematics, we see the underlying patterns and designs. Each little node in this web is intimately related to the others around it. It is this intricate web of concepts that we need to pay attention to. Expertise and love for the subject is directly related to the quality of our comprehension. Our confidence to deal with issues related to any domain of knowledge is related to the quality of comprehension. So, we need to pay attention to the details. Taking notes is a good way to demonstrate our understanding and reinforce our learnings.

Problem Solving: The key to problem solving is practice. Math is not a spectator sport. There are no brownie points for being armchair diplomats. We need to be prepared to jump in and solve the problems that we come across. With practice and only with practice do we gain the expertise to deploy the right ammunition to crack a problem.

Goal Clarity: Solving problems in order to verify our conceptual understanding is extremely important. Most of us believe arriving at the final answer is the ultimate goal. We have come across several books on the subject, where the authors have skipped several steps and simply used the phrase "it follows from the fundamental principles ..." and made a conclusion. We disagree with this approach. The purpose of the problem solving is build the path to the solution using first principles or well-known formulas - and build an airtight reasoning on how the problem solving process moves towards the final answer. This serves as a demonstration of our understanding of the subject - basics, formulas and methods of manipulation.

We may now proceed to the topics in High School Algebra. Have fun along the way.

3 Ratio

1. Ratio represents relationship between two quantities of the same kind with respect to their magnitudes

2. It denotes how many times one quantity is contained in another.

3. If there are two quantities a and b, the relationship between them can be expressed as a ratio $a:b$ read as "a is to b".

4. The ratio $a:b$ can also be expressed as a fraction $\dfrac{a}{b}$

5. In order to compare two quantities, they must be expressed in terms of the same unit.

6. In the ratio $a:b$ or $\dfrac{a}{b}$, a is called the antecedent or the numerator and b is called the consequent or denominator.

7. The value of ratio remains unaltered if the antecedent and the consequent are multiplied or divided by the same quantity.

8. Since the $a:b$ is the same as $\dfrac{a}{b}$, all laws of fractions apply to ratios as well.

9. Multiplying the antecedent and consequent by the same quantity, does not impact the impact the ratio. $\dfrac{a}{b} = \dfrac{m \times a}{m \times b}$,
    $\therefore a:b = ma:mb$

10. When the two terms of ratio are interchanged, we get a second ratio which is called "The inverse ratio" of the first.

11. Inverse ratios are also known as Reciprocal ratios. $b:a$ is the reciprocal of $a:b$. This follows from the fact that $\dfrac{b}{a}$ is the reciprocal of $\dfrac{a}{b}$.

12. Two or more ratios may be compared by reducing their equivalent fractions to a common denominator. Thus suppose $a:b$ and $x:y$ are two ratios:

$$\frac{a}{b} = \frac{ay}{by} \text{ and } \frac{x}{y} = \frac{bx}{by}$$

13. Hence the ratio $a:b$ is greater than, equal to, or less than the ratio $x:y$ according as ay is greater than equal to, or less than bx.

14. The ratio of two fractions can be expressed as a ratio of two integers. Thus the ratio $\frac{a}{b}:\frac{c}{d}$ is measured by the fraction $\dfrac{\frac{a}{b}}{\frac{c}{d}}$, or $\frac{ad}{bc}$ and is therefore equivalent to the ratio $ad:bc$

15. If the ratio of any two quantities can be represented by two integers, the ratio is said to be commensurable. For example, $2:3$ is commensurable.

16. If the ratio of any two quantities cannot be represented by two integers, then the ratio is said to be incommensurable. $\sqrt{3}:\sqrt{2}$ is an example of such a ratio.

17. When antecedents and consequents of two ratios are multiplied, this results in a compounded ratio. If $a:b$ and $c:d$ are two ratio, then $ac:bd$ is the compounded ratio.

18. If the antecedent and consequent of a ratio are squared, we get the duplicate ratio of the original ratio. Therefore, $a^2:b^2$ is the duplicate ratio of $a:b$.

19. Similarly, $a^3:b^3$ is the triplicate ratio of $a:b$.

20. And, $\sqrt{a}:\sqrt{b}$ is the sub-duplicate ratio of $a:b$.

21. Let $a:b$ be the initial ratio. Let us now add x to the antecedent and consequent of the ratio. We get $a+x:b+x$. These can be represented as fractions $\dfrac{a}{b}$ and $\dfrac{a+x}{b+x}$.

    i. $\dfrac{a}{b}-\dfrac{a+x}{b+x}=\dfrac{ab+ax-ab-bx}{b(b+x)}=\dfrac{x(a-b)}{b(b+x)}$

    ii.   If $a>b$, then $\dfrac{a}{b}>\dfrac{a+x}{b+x}$

    iii.   If $a<b$, then $\dfrac{a}{b}<\dfrac{a+x}{b+x}$

    iv.   iii. $a=b$, then $\dfrac{a}{b}=\dfrac{a+x}{b+x}$

    v.   We can see that the ratio of greater inequality is diminished when we add the same quantity to both antecedent and consequent.

    vi.   The ratio of greater inequality is increased if we take the same quantity away from the antecedent and consequent.

22. If $\dfrac{a}{b}=\dfrac{c}{d}$, each of these are equal to $\dfrac{pa+qc}{pb+qd}$, where p, q are any quantities whatever.

23. As a consequence of the derivation above, it follows that when $p=q=1$, we see that $\dfrac{a}{b}=\dfrac{c}{d}=\dfrac{a+c}{b+d}$. We can restate this finding as follows. When a series of fractions are equal, each of them is equal to the sum of all the numerators divided by the sum of all the denominators.

24. Theorem:- $\dfrac{a1}{b1}, \dfrac{a2}{b2}, \dfrac{a3}{b3}$ be unequal fractions, of which the denominators are all of the same sign, then, the fraction

$$\frac{a_1 + a_2 + a_3 + \ldots a_n}{b_1 + b_2 + b_3 + \ldots b_n}$$

lies in magnitude between the greatest and least of them..

25. Rule of cross Multiplication: Consider two simultaneous equations with three unknowns: $a_1 x + b_1 y + c_1 z = 0$ and $a_2 x + b_2 y + c_2 z = 0$

   i. Then, $b_1 c_2 - b_2 c_1, c_1 a_2 - c_2 a_1, a_1 b_2 - a_2 b_1$ are proportional to x, y, z respectively.

   ii. Therefore $\dfrac{x}{b_1 c_2 - b_2 c_1} = \dfrac{y}{c_1 a_2 - c_2 a_1} = \dfrac{z}{a_1 b_2 - d_2 b_1}$

3.1 Solved problems

1. If $\dfrac{2y + 2z - x}{a} = \dfrac{2z + 2x - y}{b} = \dfrac{2x + 2y - z}{c}$ show that

$$\frac{x}{2b + 2c - a} = \frac{y}{2c + 2a - b} = \frac{z}{2a + 2b - c}$$

We know that-

$$\left(\frac{a}{b} = \frac{c}{d} = \frac{e}{f} \right) = \left(\frac{pa^n + qc^n + re^n}{pb^n + qd^n + rf^n} \right)^{1/n}$$

Set $n = 1, \dfrac{pa + qc + re}{pb + qd + rf}$

For the ratios in question (look at the denominator) choose $2, 2, -1$

Let all of them $= k$

$$\Rightarrow \frac{2(2y+2z-x)+2(2z+2x-y)-1(2x+2y-z)}{2(a)+2(b)-(c)}=k$$

$$\Rightarrow \frac{4y+4z-2x+4z+4x-2y-2x-2y+z}{2a+2b-c}=k$$

$$\Rightarrow (\text{Cancelling})\frac{9z}{2a+2b-c}=k$$

Or $\dfrac{z}{2a+2b-c}=\dfrac{k}{9}$

If we choose the co-efficient as $2,-1,2$ we get

$$\frac{2(2y+2z-x)-1(2z+2x-y)+2(2x+2y-z)}{2(a)-1(b)+2(c)}=k$$

Simplifying, $\dfrac{y}{2a+2c-b}=\dfrac{k}{9}$

Simplifying, with coefficients of $-1,2,2$

We get: $\dfrac{x}{2b+2c-a}=\dfrac{k}{9}$

Hence, all three quantities are equal to $\dfrac{k}{9}$ and hence equal.

2. **If** $(a^2+b^2+c^2)(x^2+y^2+z^2)=(ax+by+cz)^2$ **show that**
$x:a=y:b=z:c$.

$$\left(a^2+b^2+c^2\right)\left(x^2+y^2+z^2\right)=(ax+by+cz)^2$$

Expanding:

$$a^2\left(x^2+y^2+z^2\right)+b^2(x^2+y^2+z^2)+c^2(x^2+y^2+z^2)$$

$$=ax(ax+by+cz)+by(ax+by+cz)+cz(ax+by+cz)$$

$$a^2 x^2 + a^2 y^2 + a^2 z^2 + b^2 x^2 + \cancel{b^2 y^2} + b^2 z^2$$
$$+ c^2 x^2 + c^2 y^2 + \cancel{c^2 z^2}$$
$$= \cancel{a^2 x^2} + abxy + acxz + \cancel{b^2 y^2} + abxy + bcyz$$
$$+ acxz + bcyz + \cancel{c^2 z^2}$$

Cancelling; and re arranging

$$a^2 y^2 + b^2 x^2 + a^2 z^2 + c^2 x^2 + b^2 z^2 + c^2 y^2$$
$$= 2abxy + 2acxz + 2bcyz$$

Or

$$(a^2 y^2 - 2abxy + b^2 x^2) + (a^2 z^2 - 2acxz + c^2 x^2)$$
$$+ (b^2 z^2 + c^2 y^2 - 2bcyz) = 0$$

Each term is a square,

$$(ay - bx)^2 + (az - cx)^2 + (bz - cy)^2 = 0$$

Since square cannot be negative, each term must be equal to 0

$$\Rightarrow ay - bx = 0; az - cx = 0; bz - cy = 0$$

Or $ay = bx; az = cx; bz = cy$

Or $\dfrac{x}{a} = \dfrac{y}{b}; \dfrac{z}{c} = \dfrac{x}{a}; \dfrac{z}{c} = \dfrac{y}{b}$

Or $\dfrac{x}{a} = \dfrac{y}{b} = \dfrac{z}{c}$

3. **If** $l(my + nz - lx) = m(nz + lx - my) = n(lx + my - nz)$, **prove that**

$$\frac{y + z - x}{l} = \frac{z + x - y}{m} = \frac{x + y - z}{n}$$

Let $l(lx + my + nz) = k$ (1)

$$m(lx - my + nz) = k \quad (2)$$
$$n(lx + my - nz) = k \quad (3)$$

Noticing that coefficients of x, y, z inside the brackets are the same, we can try eliminating terms

Multiply (2) by "l n" and (3) by "l m" and add

$$l\,nm(l\,n - my + nz) + lmn(l\,n + my - nz) = l\,nk + lmk$$

Or $lmn[l\,n - my + nz + lx + my - nz] = kl(m+n)$

Or $lmn(2lx) = kl(m+n)$

Or $\dfrac{x}{m+n} = \dfrac{k}{2lmn}$

Similarly, multiply (1) by mn, (2) by ln and add,

$$lmn[-lx + my + nz] + l\,nm[lx - my + nz] = mnk + l\,nk$$

$$lmn(2nz) = nk(l+m)$$

$$\frac{z}{l+m} = \frac{k}{2lmn}$$

Next, multiply (1) by mn, (3) by lm and add

$$mnl[-lx + my + nz] + lmn[lx + my - nz] = mnk + lmk$$

$$lmn(2my) = mk(l+n)$$

$$\frac{y}{l+n} = \frac{k}{2lmn}$$

So, $\dfrac{x}{m+n} = \dfrac{y}{l+n} = \dfrac{z}{l+m} = \dfrac{k}{2lmn} = q$, say

Adding them with co-efficient $1, 1, -1$

$$q = \frac{x+y-z}{m+n+l+n-(l+m)} = \frac{x+y-z}{2n}$$

Or $\dfrac{x+y-z}{n} = \dfrac{q}{2}$

Adding with co-efficient $1, -1, 1$

$$q = \frac{x - y + z}{m + n - (l + n) + l + m} = \frac{x - y + z}{2m}$$

Or $\dfrac{x - y + z}{x} = \dfrac{q}{2}$

Adding with co-efficient $-1, 1, 1$

$$q = \frac{-x + y + z}{-(m + n) + l + n + l + m} = \frac{y + z - x}{2l}$$

Or $\dfrac{y + z - x}{l} = \dfrac{q}{2}$

Hence $\dfrac{x + y - z}{n} = \dfrac{x + z - y}{m} = \dfrac{y + z - x}{l}$

4. **Show that the eliminant of** $ax + cy + bz = 0, cx + by + az = 0,$
    $bx + ay + cz = 0$ **is** $a^3 + b^3 + c^3 - 3abc = 0$

    $$ax + cy + bz = 0 \quad (1)$$
    $$cx + by + az = 0 \quad (2)$$
    $$bx + ay + cz = 0 \quad (3)$$

Cross multiply $\left. \begin{matrix} b & a & c & b \\ a & c & b & a \end{matrix} \right\}$ applied to (2) and (3)

$$\frac{x}{bc - a^2} = \frac{y}{ab - c^2} = \frac{z}{ca - b^2} = k$$

Substitute in (1)

$$ak(bc - a^2) + ck(ab - c^2) + bk(ac - b^2) = 0$$

Divide by k,

$$3abc - a^3 - b^3 - c^3 = 0$$

Or $a^3 + b^3 + c^3 - 3abc = 0$ (multiply by -1)

This is the eliminant.

5. Eliminate x, y, z from the equations

$ax + by + gz = 0, hx + by + fz = 0, \; gx + fy + cz = 0.$

$$ax + by + qz = 0$$

$$hx + by + fz = 0$$

$$gx + fy + cz = 0$$

$$\left. \begin{matrix} b & f & h & b \\ f & c & g & f \end{matrix} \right|$$

Cross multiply

$$\frac{x}{bc - fz} = \frac{y}{fg - hc} = \frac{z}{hf - bg} = k$$

Or $n = k(bc - f^2); \; y = k(fg - hc); z = k(hf - bg)$

Substituting in (2), dividing out k

$$a(bc - f^2) + h(fg - hc) + g(hf - bg) = 0$$

$$abc - af^2 + fgh - ch^2 + fgh - bg^2 = 0$$

$$abc + 2fgh - af^2 - bg^2 - ch^2 = 0 \text{ is the eliminant.}$$

6. If $x = cy + bz, y = az + cx, z = bx + ay,$ **show that**

$$\frac{x^2}{1 - a^2} = \frac{y^2}{1 - b^2} = \frac{z^2}{1 - c^2}$$

$$x - cy - bz = 0 \quad (1)$$

$$cx - y + az = 0 \quad (2)$$

$$bx + ay - z = 0 \quad (3)$$

$$\left\{ \begin{matrix} -c & -b & 1 & -c \\ -1 & a & c & -1 \end{matrix} \right]$$

Cross-multiply 1 and 2

$$\frac{x}{-ac-b} = \frac{y}{-bc-a} = \frac{z}{-1+c^2} = k$$

Or $x = (b+ac)k_1, y = (a+bc)k_1, z = (1-c^2)k_1 \ (k_1 = -k)$

Substitute in (3)

$$b(b+ac) + a(a+bc) - (1-c^2) = 0$$

$$a^2 + b^2 + 2abc - 1 + c^2 = 0$$

Similarly $\dfrac{x^2}{(b+ac)^2} = \dfrac{y^2}{(a+bc)^2} = \dfrac{z^2}{(1-c^2)^2}$

Or $\dfrac{x^2}{b^2 + 2abc + a^2c^2} + = \dfrac{y^2}{a^2 + 2abc + b^2c^2} = \dfrac{z^2}{(1-c^2)^2}$

Or $\dfrac{x^2}{1-c^2 - a^2 + a^2c^2} = \dfrac{y^2}{1-c^2 - b^2 + b^2c^2} = \dfrac{z^2}{(1-c^2)^2}$

$$\frac{x^2}{(1-a^2)(1-c^2)} = \frac{y^2}{(1-c^2)(1-c^2)} = \frac{z^2}{(1-c^2)^2}$$

Or $\dfrac{x^2}{1-a^2} = \dfrac{y^2}{1-b^2} = \dfrac{z^2}{1-c^2}$

7. **Given that** $a(y+z) = x, b(z+x) = y, c(x+y) = z,$ **prove that**
$bc + ca + ab + 2abc = 1.$

$$-x + ay + az = 0 \ (1)$$

$$bx - y + bz = 0 \ (2)$$

$$cx + cy - z = 0 \ (3)$$

$$\begin{Bmatrix} -1 & b & b & -1 \\ c & -1 & c & c \end{Bmatrix}$$

Cross – multiply 2 and 3

$$\Rightarrow \frac{x}{1-bc} = \frac{y}{bc+b} = \frac{z}{bc+c} = k$$

$$\Rightarrow -k(1-bc) + ak(bc+b) + ak(bc+c) = 0$$

Substitute in (1)

$$\Rightarrow -1 + bc + abc + ab + abc + ac = 0 \quad \text{(Remove k)}$$

Or $2abc + bc + ca + ab = 1$ is the eliminate

8. **Solve:** $3x - 4y + 7z = 0, 2x - y - 2z = 0, 3x^3 - y^3 + z^3 = 18.$

$$3x - 4y + 7z = 0$$

$$2x - y - 2z = 0$$

$$3x^3 - y^3 + z^3 = 18$$

$$\begin{cases} -4 & 7 & 3 & -4 \\ -1 & -2 & 2 & -1 \end{cases},$$

Cross multiply

$$\frac{x}{8+7} = \frac{y}{14+6} = \frac{z}{-3+8} \quad \text{Or} \quad \frac{x}{15} = \frac{y}{20} = \frac{z}{5}$$

$$\frac{x}{3} = \frac{y}{4} = z$$

$$x = 3z; y = 4z$$

Substituting $3 \cdot (3z)^3 - (4z)^3 + z^3 = 18$

$$z^3 18 = 18 \Rightarrow z^3 = 1 \Rightarrow z = 1$$

$$x = 3, y = 4; z = 1$$

9. **Solve:** $x + y = z, 3x - 2y + 17z = 0, x^3 + 3y^3 + 2z^3 = 167$

$$x + y - z = 0 \quad (1)$$

$$3x - 2y + 17z = 0 \quad (2)$$

$$x^3 + 3y^3 + 2z^3 = 167 \quad (3)$$

$$\begin{Bmatrix} 1 & -1 & 1 & 1 \\ -2 & 17 & 3 & -2 \end{Bmatrix} \text{ (2) and (3) cross-multiply}$$

$$\frac{x}{17-2} = \frac{y}{-3-17} = \frac{z}{-2-3} \text{ or } \frac{x}{15} = \frac{y}{-20} = \frac{z}{-5}$$

Or $x = -3z;\ y = 4z$

Substituting: $(-3z)^3 + 3(4z)^3 + 2z^3 = 167$

$$(-27 + 3 \cdot 64 + 2)z^3 = 167$$

$$(194 - 27)z^3 = 167 \Rightarrow 167z^3 = 167$$

$$z^3 = 1 \Rightarrow z = 1$$

$$x = -3,\ y = 4$$

10. Solve: $7yz + 3zx = 4xy, 21yz - 3zx = 4xy, x + 2y + 3z = 19.$

$$\begin{Bmatrix} 3 & -4 & 7 & 3 \\ -3 & -4 & 21 & -3 \end{Bmatrix};$$

Cross–multiply the first two equations

$$\frac{yz}{-12-12} = \frac{zx}{-84+28} = \frac{xy}{-21-63} \quad \frac{yz}{-24} = \frac{zx}{-56} = \frac{xy}{-84}$$

Multiply by -4; divide by xyz

$$\frac{1}{6x} = \frac{1}{14y} = \frac{1}{21z} \text{ Or } 6x = 14y = 21z$$

Or: $x = \dfrac{7}{2}z;\ y = \dfrac{3}{2}z$

Substituting: $\dfrac{7}{2}z + 2\left(\dfrac{3}{2}z\right) + 3z = 19 \Rightarrow \dfrac{19}{2}z = 19 \Rightarrow z = 2$

$$z = 2; x = 7; y = 3$$

4 Proportion

1. When $a:b=c:d$, then a,b,c and d are said to be in proportion.

2. This fact is represented as $a:b=c:d$ or $a:b::c:d$. This is read as "a is to b is as is to c is to d"

3. When $a:b::c:d$, then a and d are called extremes and b and c are called means.

4. $a:b::c:d \Rightarrow \dfrac{a}{b} = \dfrac{c}{d} \Rightarrow ad = bc$

5. In other words, product of means is equal to product of extremes, when four quantities are in proportion.

6. $a,b,c,d\ldots$ are said to be in continued proportion if $a:b=b:c=c:d=\ldots$.

7. If $a:b=b:c$ then a is called the first proportional, b is called mean proportional and c is called the third proportional.

8. If $a:b=b:c$, then $b^2 = ac$. In other words, if three quanities are in continued proportion, then the first is to third is the duplicate ratio of first is to second.

9. If $a:b=c:d$ and $e:f=g:h$, then $ae:df=cg:dh$ If $\dfrac{a}{b} = \dfrac{c}{d}$ and

   $\dfrac{e}{f} = \dfrac{g}{h}$, then $\dfrac{ae}{bf} = \dfrac{cg}{dh}$. This implies $ae:df=cg:dh$

10. If $a:b=c:d$ and $b:x=d:y$. then $a:x=c:y$

11. Rules of proportion

    i. Invertedo Rule: If two ratios are equal, their inverses will also be equal.

    $$\text{If } \frac{a}{b} = \frac{c}{d}, \text{then} \frac{b}{a} = \frac{d}{c}$$

    ii. Alternendo rule: If two ratios are equal, the ratios formed by the first and third terms will be equal to ratio of the second terms to the fourth terms.

If $\dfrac{a}{b} = \dfrac{c}{d}, \dfrac{a}{c} = \dfrac{b}{d}$

iii. Componendo rule: If two ratios are equal, then the ratios derived by adding units to the original ratios, are also equal. If $\dfrac{a}{b} = \dfrac{c}{d}$ by adding 1 to both sides $\dfrac{a}{b} + 1 = \dfrac{c}{d} + 1$

$$\dfrac{a+b}{b} = \dfrac{c+d}{d}$$

iv. Dividendo rule: If two ratios are equal, then the ratios derived by subtracting unit from both the original ratios, are also equal. Ex: If $\dfrac{a}{b} = \dfrac{c}{d}$ by subtracting 1.

$$\dfrac{a}{b} - 1 = \dfrac{c}{d} - 1 \quad \dfrac{a-b}{b} = \dfrac{c-d}{d}$$

v. Componendo et Dividendo Rule :

If $\dfrac{a}{b} = \dfrac{c}{d}$

by componendo rule

$$\dfrac{a+b}{b} = \dfrac{c+d}{d}$$

By Dividendo rule

$$\dfrac{a-b}{b} = \dfrac{c-d}{d}$$

Divide (1) by (2), we get

$$\dfrac{\dfrac{a+b}{b}}{\dfrac{a-b}{b}} = \dfrac{\dfrac{c+d}{d}}{\dfrac{c-d}{d}}$$

$$\dfrac{a+b}{a-b} = \dfrac{c+d}{c-d}$$

This is called componendo et dividendo rule.

4.1 Solved problems:

1. **Find the fourth proportional to 3, 5, 27.**

   3, 5, 27,?,   say n

   $$\frac{3}{5} = \frac{27}{n} \quad \text{Or} \quad n = \frac{37 \times 5}{3} = 45$$

2. **Find the mean proportional between (i) 6 and 24 (ii)** $360a^4$ **and** $250\,a^2b^2$ **.**

   (i) 6 and 24

   $$6 \text{ and } 24 = \sqrt{6 \times 24} = \sqrt{144} = 12$$

   (ii) 360 a^4 **and 250** a^2b^2

   $$360a^4, \; 250a^2b^2 = \sqrt{360 \cdot a^4 \times 250a^2b^2}$$

   $$= \sqrt{36 \times 25 \times 100 \times a^4 \cdot a^2b^2}$$

   $$= 6 \times 5 \times 10 \times a^2 \times ab$$

   $$= 300a^3b$$

3. **Find the third proportional to** $\dfrac{x}{y} + \dfrac{y}{x}$ **and** $\dfrac{x}{y}$

   $$\frac{x}{y} + \frac{y}{x} \text{ and } \frac{x}{y};$$

   Let the third proportional be P

$$p = \dfrac{\left(\dfrac{x}{y}\right)^2}{\dfrac{x}{y}+\dfrac{y}{x}} = \dfrac{x^2}{y^2\left(\dfrac{x^2+y^2}{xy}\right)} = \dfrac{x^3}{y\left(x^2+y^2\right)}$$

If a: b = c: d, prove that

4. $a^2c + ac^2 : b^2d + bd^2 = (a+c)^3 : (b+d)^3$

$$\dfrac{a^2c+ac^2}{b^2d+bd^2} = \dfrac{(a+c)^3}{(b+d)^3}$$

Let $\dfrac{a}{b} = \dfrac{c}{d} = k \qquad \Rightarrow a = bk; c = dk$

$$\dfrac{a^2c+ac^2}{b^2d+bd^2} = \dfrac{b^2k^2 \cdot dk + bk \cdot d^2k^2}{b^2b+bd^2} = k^3$$

$$\dfrac{(a+c)^3}{(b+d)^3} = \dfrac{(bk+dk)^3}{(b+d)^3} = k^3$$

5. $pa^2 + qb^2 : pa^2 - qb^2 = pc^2 + qd^2 : pc^2 - qd^2$

$$\dfrac{a}{b} = \dfrac{c}{d} \Rightarrow \dfrac{a^2}{b^2} = \dfrac{c^2}{d^2} \Rightarrow \dfrac{pa^2}{qb^2} = \dfrac{pc^2}{qd^2}$$

$$\dfrac{pa^2+qb^2}{pa^2-qb^2} = \dfrac{pc^2+qd^2}{pc^2-qd^2}$$

6. $a - c : b - d = \sqrt{a^2+c^2} : \sqrt{b^2+d^2}$

$a = kb, c = kd$

$$\dfrac{a-c}{b-d} = \dfrac{kb-kd}{b-d} = k$$

$$\frac{\sqrt{a^2+c^2}}{\sqrt{b^2+d^2}} = \frac{\sqrt{k^2b^2+k^2d^2}}{\sqrt{b^2+d^2}} = \frac{\sqrt{k^2(b^2+d^2)}}{\sqrt{b^2+d^2}} = k$$

7. $\sqrt{a^2+c^2} : \sqrt{b^2+a^2} = \sqrt{ac+\dfrac{c^3}{a}} : \sqrt{bd+\dfrac{d^3}{b}}$

$$\frac{\sqrt{a^2+c^2}}{\sqrt{b^2+d^2}} = \frac{\sqrt{k^2b^2+k^2d^2}}{\sqrt{b^2+d^2}} = k$$

$$\frac{\sqrt{ac+\dfrac{c^3}{a}}}{\sqrt{bd+\dfrac{d^3}{b}}} = \frac{\sqrt{k^2bd+\dfrac{k^3d^3}{kb}}}{\sqrt{bd+\dfrac{d^3}{b}}} = \frac{k\sqrt{bd+\dfrac{d^3}{b}}}{\sqrt{bd+\dfrac{d^3}{b}}} = k$$

If a, b, c, d are in continued proportion, prove that

8. $a : b+d = c^3 : c^2d+d^3$

$$\frac{a}{b} = \frac{b}{c} = \frac{c}{d} = k$$

$$c = dk; b = ck = dk^2 ; a = bk = dk^3$$

$$\frac{a}{b+d} = \frac{dk^3}{dk^2+d} = \frac{k^3}{k^2+1}$$

$$\frac{c^3}{c^2d+d^3} = \frac{d^3k^3}{d^3k^2+d^3} = \frac{k^3}{k^2+1}$$

9. $2a+3d : 3a-4d = 2a^3+3b^3 : 3a^3-4b^3$

$$\frac{2a+3d}{3a-4d} = \frac{2k^3d+3d}{3k^3d-4d} = \frac{2k^3+3}{3k^3-4}$$

$$\frac{2a^3 + 3b^3}{3a^3 - 4b^3} = \frac{2(k^3 d) + 3(k^2 d)^3}{3(k^3 d)^3 - 4(k^2 d)^3} = \frac{2k^9 d^3 + 3k^6 d^3}{3k^9 d^3 - 4k^6 d^3}$$

$$= \frac{k^6 d^3}{k^6 d^3} \left[\frac{2k^3 + 3}{3k^3 - 4} \right] = \frac{2k^3 + 3}{3k^3 - 4}$$

10. $(a^2 + b^2 + c^2)(b^2 + c^2 + d^2) = (ab + bc + cd)^2$

$$(a^2 + b^2 + c^2)(b^2 + c^2 + d^2)$$

$$= (k^6 d^2 + k^4 d^2 + k^2 d^2)(k^4 d^2 + d^2 k^2 + d^2)$$

$$= (k^5 d^2 + k^3 d^2 + kd^2)(k^5 d^2 + k^3 d^2 + kd^2)$$

$$= (k^5 d^2 + k^3 d^2 + kd^2)^2$$

$$= (k^3 d \cdot k^2 d + k^2 d \cdot kd + kd \cdot d)^2$$

$$= (ab + bc + cd)^2$$

5 Variation

1. We will now explore the relationship between two quantities. The relationship of interest revolves around a key question. How does one quantity vary with respect to another ? If they do vary, what is kind of proportion do they vary in ?

2. There are three kinds of variation

      i.   Direct Variation or Direct Proportion

      ii.   Inverse Variation or Inverse Proportion

      iii.   Joint Variation

3. We will now take a close look at each of these terms.

4. Direct Variation

      i.   When a quantity a varies directly as another quantity b, when we observe a change in a results in a change in b in the same ratio.

      ii.   We say a varies as b or a is directly proportional to b

      iii.   This is written as $a \propto b$

      iv. $a \propto b \dfrac{a}{b} = k$ or $a = k \times b$ where k is a constant.

5. Inverse Variation

      i.   When a quantity a varies inversely as b, then a varies directly as reciprocal of b.

      ii.   Thus $a \propto \dfrac{1}{b} a = k \times \dfrac{1}{b} a \times b = k$, where k is a constant.

6. Joint Variation

      i.   (a) a is said to vary jointly as b, c, when a varies as the product of b and c.

      ii.   (b) $a \propto b \times c$

      iii.   (c) $a = k \times bc$, where k is a constant

    iv.   (d) If a varies as b when c is constant, and a varies as c when b is constant, then a is said to vary jointly as b and c.

7.   Real World Scenarios of variation

    i.   Time and Work: Time to complete a job varies directly as the amount of work and inversely as the number of work men employed.

    ii.   Time and Distance: Time taken to travel a distance varies directly as the distance but varies inversely as time.

8.   Properties of variation

    i.   If $a \propto b b \propto a$

    ii.   If $a \propto b$ and $b \propto c$, then $a \propto c$

    iii.   If $a \propto b$ and $b \propto c$, then

         a. $(a+b) \propto c$

         b. $(a-b) \propto c$

         c. $\sqrt{ab} \propto c$

    iv.   If $a \propto bc$, then $b \propto \dfrac{a}{c}$ and $c \propto \dfrac{a}{b}$

    v.   If $a \propto b$ then $a^n \propto b^n$

    vi.   If $a \propto bc$ and If $b \propto c$ a c is a constant.

    vii.   If $a \propto b$ and $a \propto \dfrac{1}{c}$, then $a \propto \dfrac{b}{c}$

    viii.   If $a \propto b$ and $b \propto \dfrac{1}{c}$, then $a \propto \dfrac{1}{c}$

5.1   Solved problems

1. **If x varies as y, and** $x = 8$ **when** $y = 15$, **find x when** $y = 10$.

$x \propto y \Rightarrow x = ky$

When $x = 8, y = 15$

We substitute the values in the equation and get

$$8 = k \times 15; k = \frac{8}{15}$$

$$x = \frac{8}{15} \times 10 = \frac{16}{3}$$

2. **If the square of x varies as the cube of y and** $x = 3$ **when** $y = 4,$ **find the value of y when** $x = \dfrac{1}{\sqrt{3}}$

$$x^2 \propto y^3 \Rightarrow x^2 = ky^3 \Rightarrow k = \frac{x^2}{y^3}$$

$$x = 3, y = 4 \Rightarrow k = \frac{3^2}{4^3} = \frac{9}{64}$$

$$x = \frac{1}{\sqrt{3}} \Rightarrow y = \left(\frac{x^2}{k}\right)^{\frac{1}{3}} = \left(\left(\frac{1}{\sqrt{3}}\right)^2 \cdot \frac{64}{9}\right)^{\frac{1}{3}}$$

$$= \left(\frac{64}{3 \cdot 9}\right)^{\frac{1}{3}} = \frac{4}{3}$$

3. **If A varies as C, and B varies as C, then prove that** $A \pm B$ **and** $\sqrt{AB}$ **will each vary as C.**

$A \times C \Rightarrow A = kC$

$B \times C \Rightarrow B = lC$

$A \pm B = kC \pm lc = C(k \pm l) \Rightarrow (A \pm B) \propto C$

$\sqrt{AB} = \sqrt{kc \cdot lc} = \sqrt{kl} \cdot C \Rightarrow \sqrt{AB} \propto C$

4. P varies directly as Q and inversely as R; also $P = \dfrac{2}{3}$ when $Q = \dfrac{3}{7}$ and $R = \dfrac{9}{14}$; Find Q when $P = \sqrt{48}$ and $R = \sqrt{75}$.

$$P \times \frac{Q}{R} \Rightarrow P = k \cdot \frac{Q}{R}$$

$$P = \frac{2}{3}, Q = \frac{3}{7}, R = \frac{9}{14} \Rightarrow k = \frac{RP}{Q} = \frac{\frac{9}{14} \cdot \frac{2}{3}}{\frac{3}{7}}$$

$$k = \frac{18 \times 7}{14 \times 3 \times 3} = 1$$

$$Q = \frac{R \cdot P}{k} = \sqrt{45} \cdot \sqrt{75} = \sqrt{16 \cdot 3 \cdot 25 \cdot 3}$$

$$= 4 \cdot 3 \cdot 5 = 60$$

5. If y varies as the sum of two quantities, of which one varies directly as x and the other inversely as x; and if $y = 6$ when $x = 4$ and $y = 3\dfrac{1}{3}$ when $x = 3$; find the equation between x and y

Let $y = y_1 + y_2$

$$y_1 \times x \Rightarrow y_1 = kx$$

$$y_2 \times \frac{1}{x} \Rightarrow y_2 = \frac{l}{x}$$

Or $y = kx + \dfrac{l}{x}$

$$x = 4, y = 6 \Rightarrow 6 = k \cdot 4 + \frac{l}{4}$$

Or $16k + l = 24$

$$y = 3\frac{1}{3}, x = 3 \Rightarrow 3\frac{1}{3} = k \cdot 3 + \frac{l}{3} = \frac{10}{3}$$

$$\Rightarrow 9k + l = 10$$

Subtracting; $7k = 14 \Rightarrow k = 2$

$$l = -8$$

$$y = 2x - \frac{8}{x}$$

6. **If A varies directly as the square root of B and inversely as the cube of C, and if** $A = 3$ **when** $B = 256$, **and** $C = 2$, **find B when** $A = 24$ **and** $C = \dfrac{1}{2}$

$$A \propto \frac{\sqrt{B}}{C^3} = \frac{k\sqrt{B}}{C^3} \Rightarrow k = \frac{AC^3}{\sqrt{B}}$$

$$A = 3, B = 256, C = 2 \Rightarrow k = \frac{3 \cdot 2^3}{\sqrt{256}} = \frac{3 \cdot 8}{16} = \frac{3}{2}$$

$$A = 24, C = \frac{1}{2} \Rightarrow \sqrt{B} = \frac{AC^3}{1C} \Rightarrow B = \left(\frac{AC^3}{k}\right)^2$$

$$B = \left(\frac{24 \cdot \left(\frac{1}{2}\right)^3}{\left(\frac{3}{2}\right)}\right)^2 = \left(\frac{\frac{24}{8}}{\left(\frac{3}{2}\right)}\right)^2 = 2^2 = 4$$

7. **Given that** $x+y$ **varies as** $z+\dfrac{1}{z}$, **and that** $x-y$ **varies as** $z-\dfrac{1}{z}$, **find the relation between x and z, provided that** $z=2$ **when** $x=3$ **and** $y=1.$ $x+y=k\left(z+\dfrac{1}{z}\right)$

$$x-y=l\left(z-\dfrac{1}{z}\right)$$

$$x=3, y=1, z=2$$

$$3+1=k\left(2+\dfrac{1}{2}\right)\Rightarrow k=\dfrac{4}{\left(\dfrac{5}{2}\right)}=\dfrac{8}{5}$$

$$3-1=l\left(2-\dfrac{1}{2}\right)\Rightarrow l=\dfrac{2}{\left(\dfrac{3}{2}\right)}=\dfrac{4}{3}$$

$$(x+y)+(x-y)=k\left(z+\dfrac{1}{z}\right)+l\left(z-\dfrac{1}{z}\right)$$

$$\text{Or } x=\dfrac{\dfrac{8}{5}\left(z+\dfrac{1}{2}\right)+\dfrac{4}{3}\left(z-\dfrac{1}{z}\right)}{2}$$

$$=\dfrac{8}{10}z+\dfrac{8}{10z}+\dfrac{1}{3}z-\dfrac{2}{3z}\left|\dfrac{8}{10}+\dfrac{2}{3}\right.$$

$$=\dfrac{44}{30}z+\dfrac{4}{30z}$$

$$=\dfrac{22}{15}z+\dfrac{2}{15z}$$

8. If A varies as B and C jointly, while B varies as D^2 and C varies inversely as A, show that A varies as D.

$$A \propto BC \Rightarrow A = kBC$$

$$B \propto D^2 \Rightarrow B = lD^2$$

$$C \propto \frac{1}{A} \Rightarrow C = \frac{m}{A}$$

$$A = k \cdot B \cdot C = k \cdot lD^2 = \frac{m}{A} = lk \cdot m\frac{D^2}{A}$$

Or $A^2 = klmD^2$ or $A = \pm klmD$

Hence $A \propto D$

9. If y varies as the sum of three quantities of which the first is constant, the second varies as x and the third as x^2; and if $y = 0$ when $x = 1$, $y = 1$ when $x = 2$ and $y = 4$ when $x = 3$ find y when $x = 7$

$$y = y_1 + y_2 + y^3 ; y_1 = c; y_2 = bx; y_3 = ax^2$$

$$\Rightarrow y = ax^2 + bx + c$$

$$y = 0, x = 1 \Rightarrow 0 = a + b + c \ (1)$$

$$y = 1, x = 2 \Rightarrow 1 = 4a + 2b + c \ (2)$$

$$y = 4, x = 3 \ y \Rightarrow 4 = 9a + 3b + c \ (3)$$

Subtracting 1 from 2, 1 from 3

$$1 = 3a + b \ (4)$$

$$4 = 8a + 2b \ (5)$$

Subtracting $2 \times (4)$ from 5

$$4 - 2 \cdot 1 = 8a + 2b - 2(3a + b)$$

$$\Rightarrow 2 = 2a \text{ or } a = 1$$

$$b = 1 - 3a = -2$$

$$c = 1 - a - b = -1 + 2 = 1$$

$$y = x^2 - 2x + 1 = (x-1)^2$$

$$x = 7 \Rightarrow y = (x-1)^2 = (7-1)^2 = 6^2$$

$$= 36$$

10. When a body falls from rest its distance from the starting point varies as the square of the time it has been falling: if a body falls through 122.6m in 5 seconds, how far does it fall in 10 seconds? Also how far does it fall in the 10th second?

Distance $D \propto t^2$ time fallen

$$D = kt^2$$

Or $122 \cdot 6 = k \cdot 25$ or $k = \dfrac{122 \cdot 6}{25}$

$$k = \frac{122 \cdot 6 \times 4}{100} = 4 \cdot 904$$

$$t = 10 \Rightarrow D = k \cdot t^2 = 4 \cdot 904 \times 100$$

$$= 490 \cdot 4m$$

10th sec $= 10 \sec - 9 \sec$

$$= k(10^2 - 9^2) = 19k = 93 \cdot 176m$$

6 Arithmetic Progression

1. Quantities are said to be in arithmetic progression, when they increase or decrease by the same amount, which is also known as the common difference.

2. The difference between two successive terms in the series is called the common difference.

3. Arithmetic progression is abbreviated as AP.

4. If a is the first term in an AP, and d is the common difference, then a, $a+d$, $a+2d$, $a+3d$, ..., $a+(n-1)d$ form the first n terms in the series.

5. Clearly the general expression for the nth term in an AP is $a+(n-1)d$, where a is the first term and d is the common difference.

6. Let us now find the sum of the first n terms in an AP, whose first term is a and common difference is d.

    The sum to n terms is often represented by S_n.

    $$S_n = a+a+d+a+2d+a+3d...+a+(n-1)d$$

    Writing the sum in reverse, we have:

    $$S_n = a+(n-1)d+a+(n-2)d+a+(n-3)d+...+a$$

    Adding the above equations, we have:

    $$2S_n = 2a+(n-1)d+2a+(n-1)d+2a+(n-1)d...+a$$

    $$2S_n = n \times (2a+(n-1)d)$$

    $$S_n = \frac{n}{2}(2a+(n-1)d)$$

7. Arithmetic mean

    a. When three quantities are in Arithmetic progression the middle one is said to be the arithmetic mean of the other two.

b. If a is the first term and d is the common difference, then a, $a+d$ and $a+2d$ are the three terms.

c. $a+d$ is the arithmetic mean between a and $a+2d$

d. If AM is the arithmetic mean between a and b, then a, AM, b are in AP.

e. And, arithmetic mean $AM = \dfrac{a+b}{2}$

f. We can insert as many terms as we desire, between two quanitities a and b, such that a, t_1, t_2, t_3, ..., t_n and b are in AP.

g. The terms thus inserted are called arithmetic means.

h. If you insert n terms between a and b, then there will be $n+2$ terms in the AP.

i. The nth term of an AP $= a+(n-1)d$

j. b is the $(n+2)^{th}$ term in the AP $= a+(n+2-1)d = a+(n+1)d$

k. $d = \dfrac{b-a}{n+1}$

l. The series is a, $a+\dfrac{b-a}{n+1}$, $a+2\dfrac{b-a}{n+1}$, ...

8. Tips for handling problems in AP

a. If you are given the sum of n terms in an arithmetic progression, it may be useful to formulate your terms appropriately.

b. For example, if sum of three successive terms of an AP are given, the terms $a-d$, a and $a+d$ will aid problem solving. When you add these terms, you are left with a sum $3a$ and therefore the middle term can be determined in an instant. The point to note is that this formulation works well where the number of terms are odd.

c. This is similar to problems in elementary algebra where we choose three consecutive integers as $n-1$, n and $n+1$. In fact, the integers are in an AP with a common difference of 1.

d. If the number of terms are even, then we need a subtle change to our formulation, If we had 4 terms in the AP, then $a - \dfrac{3d}{2}, a - \dfrac{d}{2}, a + \dfrac{d}{2}, a + \dfrac{3d}{2}$ is a good formulation of the AP. This ensures that the common difference is d and the sum of the terms to be $4a$. Therefore, the common difference is simply $\dfrac{S_n - 4a}{4}$.

6.1 Solved problems

1. **Sum 2, $3\dfrac{1}{4}, 4\dfrac{1}{2}$, ... to 20 terms.**

$$d = 3\frac{1}{4} - 2 = 1\frac{1}{4} = \frac{5}{4}$$

$$\text{Sum} = \frac{n}{2}\left\{2a + (n-1)d\right\}$$

$$= \frac{20}{2}\left[2 \times 2 + (20-1)\frac{5}{4}\right] = 10\left(4 + \frac{19 \times 5}{4}\right)$$

$$= \frac{10}{4}(16 + 19 \times 5)$$

$$= 277\frac{1}{2}$$

2. **Sum 49, 44, 39, ... to 17 terms.**

$$n = 17, a = 49, d = -5$$

$$s = \frac{17}{2}(2 \times 49 + 16 \times -5) = \frac{17}{2}(98 - 80)$$

$$= 153$$

3. **Sum** $3, \dfrac{7}{3}, 1\dfrac{2}{3}, \ldots$ **to n terms.**

$$a = 3, d = \frac{-2}{3}, n = n$$

$$S = \frac{n}{2}\left(6 + (n-1)\frac{-2}{3}\right)$$

$$= \frac{n}{2}\left(\frac{18 - 2n + 2}{3}\right)$$

$$= \frac{n}{2 \times 3}(20 - 2n) = \frac{n}{3}(10 - n)$$

4. **Sum 3.75, 3.5, 3.25, ……. to 16 terms.**

$3 \cdot 75, 3 \cdot 5, 3 \cdot 25$ to 16 terms

$$n = 16, a = 3 \cdot 75, d = -0 \cdot 25$$

$$S = \frac{16}{2}(2 \times 3 \cdot 75 + 15 \times -0 \cdot 25)$$

$$= 8(7 \cdot 5 - 3 \cdot 75) = 8 \times 3 \cdot 75$$

$$= 30$$

5. **Sum** $2a - b, 4a - 3b, 6a - 5b, \ldots$ **to n terms**

$$a = 2a - b, d = 2a - 2b, n = n$$

$$S = \frac{n}{2}\big(2 \times (2a - b) + (n-1)(2a - 2b)\big)$$

$$= \frac{n}{2}\big(4a - 2b + 2n(a - b) - 2a + 2b\big)$$

$$= \frac{n}{2}\big[2n(a - b) + 2a\big] = n(n(a - b) + a)$$

$$= n^2(a-b) + an$$

6. **Sum** $\dfrac{a+b}{2}, a, \dfrac{3a-b}{2}, \ldots\ldots$ **to 21 terms**

$$a = \frac{a+b}{2}, d = \frac{a-b}{2}, n = 21$$

$$S = \frac{21}{2}\left(2 \times \left(\frac{a+b}{2}\right) + 20 \times \left(\frac{a-b}{2}\right)\right)$$

$$= \frac{21}{2}(a + b + 10a - 10b)$$

$$= \frac{21}{2}(11a - 9b)$$

7. **Insert 19 arithmetic means between** $\dfrac{1}{4}$ **and** $-9\dfrac{3}{4}$

$$n = 21, a = \frac{1}{4}, 21^{st} \text{ term} = a + 20d = -9\,{}^{3}\!/_{4}$$

$$-\frac{39}{4} = \frac{1}{4} + 20d \Rightarrow \frac{-40}{4} = 20d \Rightarrow d = -\frac{1}{2}$$

Terms are $\dfrac{1}{4}, \dfrac{-1}{4}, \dfrac{-3}{4} \ldots, -9\,{}^{1}\!/_{4}, -9\,{}^{3}\!/_{4}$

8. **Insert 17 arithmetic means between** $3\dfrac{1}{2}$ **and** $-41\dfrac{1}{2}$

$$a = \frac{7}{2}, n = 19$$

$$\text{Last} = \frac{-83}{2} = a + 18d = \frac{7}{2} + 18d$$

$$d = \frac{\dfrac{-90}{2}}{(18)} = \frac{-5}{2}$$

Means are $1, \dfrac{-3}{2} \ldots, -39$

9. **In an A.P, the first term is 2, the last term 29, the sum 155; find the difference.**

$$a = 2, \text{last} = 29, S = 155$$

$$S = \frac{n}{2}(a + l)$$

$$n = \frac{2s}{a+l} = \frac{2 \times 155}{2 + 29} = 10$$

$$d = \frac{l - a}{n - 1} = \frac{29 - 2}{10 - 1} = \frac{27}{9} = 3$$

10. **Find the sum of 35 terms of the series whose pth term is** $\dfrac{p}{7} + 2$

$$t_p = \frac{p}{7} + 2 ; a = t_1 = \frac{1}{7} + 2 = \frac{15}{7}$$

$$d = \frac{p+1}{7} + 2 - \left(\frac{p}{7} + 2\right) = \frac{1}{7}$$

$$S_{35} = \frac{35}{2}\left(2 \times \frac{15}{7} + 34 \times \frac{1}{7}\right) = \frac{35}{2 \times 7}(64) = 5 \times 32$$

$$= 160$$

11. **Given** $a = -2, d = 4$ **and** $s = 160$ **find n**

$$s = \frac{n}{2}(2a + (n-1)d)$$

$$n^2 d + (2a - d)n - 2s = 0$$

$$4n^2 + (-4 - 4)n - 320 = 0$$

$$n^2 - 2n - 80 = 0$$
$$(n-10)(n+8) = 0$$
$$n = 10 \, \text{or} -8$$

12. How many terms of the series 12, 16, 20, ... must be taken to make 208?

$$a = 12, d = 4$$
$$dn^2 + (2a - d)n - 2s = 0$$
$$4n^2 + (24 - 4)n - 416 = 0$$
$$n^2 + 5n - 104 = 0$$
$$(n+13)(n-8) = 0$$
$$n = 8 \, \text{or} -13$$

13. In an A.P. the third term is four times the first term, and the sixth term is 17; find the series.

$$t_3 = 4t_1, t_6 = 17$$
$$t_3 = a + 2d = 4a; t_0$$
$$t_6 = a + 5d = 17$$
$$\Rightarrow 3a = 2d \Rightarrow a = \frac{2}{3}d$$
$$\text{Or } \frac{2}{3}d + 5d = 17$$
$$\frac{17d}{3} = 1 > \Rightarrow d = 3$$
$$\Rightarrow a = 2$$

The series is $2, 5, 8, \cdots$

14. The 2nd, 31st and last terms of an A. P. are $7\frac{3}{4}, \frac{1}{2}$ **and respectively; find the first term and the number of terms.**

$$a + d = \frac{31}{4}; a + 30d = \frac{1}{2}; a + (n-1)d = \frac{-13}{2}$$

$$29d = \frac{-29}{4} \Rightarrow d = -\frac{1}{4} \Rightarrow a = \frac{31}{4} - d = 8$$

$$n = \frac{\left(\frac{-13}{2} - a\right)}{d} + 1 = \frac{\left(\frac{-13}{2} - \frac{16}{2}\right)}{-\frac{1}{4}} + 1 = 29 \times 2 + 1$$

$$n = 59$$

15. The sum of n terms of the series $2, 5, 8, \cdots$ **in 950.**

$$S_n = 950, a = 2, d = 3$$

$$dn^2 + (2a - d)n - 2S_n = 0$$

$$3n^2 + n - 1900 = 0$$

$$n = \frac{-1 \pm \sqrt{1 - 4 \times 3 \times -1900}}{2 \times 3}$$

$$= \frac{-1 \pm \sqrt{1 + 22800}}{6}$$

$$= \frac{-1 \pm 151}{6}$$

$$= \frac{150}{6}$$

Discarding the other solutions which are not integers, we get

$$n = 25$$

16. Sum the series $\dfrac{1}{1+\sqrt{x}}, \dfrac{1}{1-x}, \dfrac{1}{1-\sqrt{x}}, \ldots$ **to n terms.**

$$a = \frac{1}{1+\sqrt{x}}; d = \frac{1}{1-x} - \frac{1}{1+\sqrt{x}} = \frac{1+\sqrt{x}-1+x}{\left(1+\sqrt{x}\right)(1-x)}$$

$$d = \frac{x+\sqrt{x}}{\left(1+\sqrt{x}\right)(1-x)} = \frac{\sqrt{x}}{1-x}$$

$$S_n = \frac{n}{2}\left[\frac{2}{1+\sqrt{x}} + \frac{(n-1)\sqrt{x}}{1-x}\right] \quad \left| 1+\sqrt{x} = \frac{1-x}{1-\sqrt{x}} \right.$$

$$= \frac{n}{2}\left[\frac{2\left(1-\sqrt{x}\right)+(n-1)\sqrt{x}}{1-x}\right]$$

$$= \frac{n}{2}\left[\frac{2-2\sqrt{x}+n\sqrt{x}-\sqrt{x}}{1-x}\right]$$

$$= \frac{n}{2}\left[\frac{n\sqrt{x}-3\sqrt{x}+2}{1-x}\right] = \frac{n}{2}\left[\frac{n\left(\sqrt{x}-3\right)+2}{1-x}\right]$$

17. If the sum of 7 terms is 49, and the sum of 17 terms is 289, find the sum of n terms.

$$S_7 = 49 = \frac{7}{2}\left[2a+6d\right] = 7\left(a+3d\right)$$

$$S_{17} = 289 = \frac{17}{2}\left(2a+16d\right) = 17\left(a+8d\right)$$

$$a+3d = 7; a+8d = 17$$

$$5d = 10; d = 2$$

$$a = 1$$

$$S_n = \frac{n}{2}\left[2+(n-1)2\right] = n^2$$

18. **If the pth, qth, rth terms of an A.P are a, b, c respectively, show that** $(q-r)a+(r-p)b+(p-q)c=0$

$t_p = t_1 + (p-1)d = a$ (1)

$t_q = t_1 + (q-1)d = b$ (2)

$t_r = t_1 + (r-1)d = c$ (3)

$(q-r)a + (r-p)b + (p-q)c$

$= (q-r)(t_1 + (p-1)d) + (r-p)(t_1 + (q-1)d)$
$\quad + (p-q)(t_1 + (r-1)d)$

$= t_1(q-r+r-p+p-q) + d(q-r)(p-1)$
$\quad + (r-p)(q-1) + (p-q)(r-1)$

$= t_1 \times 0 + d \begin{bmatrix} p(q-r) - (q-r) + q(r-p) - (r-p) \\ + r(p-q) - (p-q) \end{bmatrix}$

$= 0 + d \begin{bmatrix} pq - pr + qr - pq + pr - rq \\ + q - r + p - q + r - p \end{bmatrix}$

$= 0$

19. **The sum of p terms of an A.P. is q, and the sum of q terms is p; find the sum of** $p+q$ **terms**

$S_p = q; \ S_q = P; \ S_{p+q} = ?$

$S_{p+q} = \left(\dfrac{p+q}{2}\right)(2a + (p+q-1)d)$

$S_p = \dfrac{P}{2}(2a + (p-1)d) = q$

$S_q = \dfrac{q}{2}(2a + (q-1)d) = P$

$S_{p+q} = \dfrac{p}{2}(2a + (p-1)d + qd) + \dfrac{q}{2}(2a + (q-1)d + pd)$

$$= q + \frac{pqd}{2} + p + \frac{pqd}{2}$$

$$= p + q + pqd$$

$S_q - S_p$ Gives

$$d = \frac{\dfrac{2p}{q} - \dfrac{2q}{p}}{q - p} = 2\left(\frac{p^2 - q^2}{(q-p)qp}\right)$$

$$= -2\frac{(p+q)}{pq}$$

$$\Rightarrow S_{p+q} = p + q - 2pq\left(\frac{p+q}{pq}\right)$$

$$= (p+q) - 2(p+q)$$

$$= -(p+q)$$

20. The sum of four integers in A.P. is 24, and their product is 945; find them.

$$a - 3d, a - d, a + d, a + 3d$$

Sum $= 4a = 24 \Rightarrow a = 6$

$$(a-d)(a-3d)(a+d)(a+3d) = 945$$

$$\Rightarrow (a^2 - d^2)(a^2 - 9d^2) = 945$$

$$(d^2 - 36)(9d^2 - 36) = 945$$

$$(d^2 - 36)(d^2 - 9) = 105$$

$$d^4 - 40d^2 + 144 = 105$$

$$d^4 - 40d^2 + 39 = 0$$

$$(d^2 - 39)(d^2 - 1) = 0$$

$$d^2 = 39 \, \text{or} \, 1$$

$$d = \pm\sqrt{39} \text{ or } \pm 1$$

Since integers, $d = \pm 1$

So, $3, 5, 7, 9$

7 Geometric Progression

1. Definition: Quantities are said to be in Geometrical Progression when they increase or decrease by a common factor (common ratio).

2. For example, $2, 4, 8, 16, 32, 64$ are in geometric progression with a common ratio of 2.

3. Again $\frac{1}{3}, \frac{1}{9}, \frac{1}{27}, \frac{1}{81}$ are in geometric progression with a common ratio of $\frac{1}{3}$.

4. The generalized representation of a geometric progression is therefore $a, ar, ar^2, ar^3, \ldots ar^n$, where a is the first term and r is the common ratio.

5. Clearly, when r is greater than unity, the successive terms become increasingly large. In such cases, the progression does not converge.

6. If n be the number of terms, and if l denote the last, or n^{th} term, we have $l = ar^{n-1}$

7. Geometrical mean:- When three quantities are in Geometrical progression, the middle one is called the geometric mean between the other two.

8. Let a, b be the two quantities, G the geometric mean, then we can see that a, G, b are in geometric progression.

   a. This means $G = \sqrt{ab}$ and

   b. $\dfrac{G}{a} = \dfrac{b}{G} = $ common ratio

9. Let us turn our attention to the task of inserting a given number of geometric means between two given quantities. Let a and b be the first and last terms of the geometric progression.

   a. If we inserted n geometric means between a and b, then the geometric progression will have $n + 2$ terms.

b. a, b and the n terms in between constitute the $n+2$ terms.

c. The last term in the GP is $b = ar^{n+1}$

d. This means $r^{n+1} = \dfrac{b}{a}$. Therefore, common ratio

$$r = \left(\frac{b}{a}\right)^{\frac{1}{n+1}}$$

e. At first blush, this looked complicated. In fact, it isn't. Let us insert 4 geometric means between 1 and 243. If we insert 4 terms in between, we have 6 terms in all. Clearly, $243 = 1 \times r^5$. This means the common ratio $r = 243^{\frac{1}{5}}$, or $r = 3$. Therefore the series is $1, 3, 9, 27, 81$ & 243. And $3, 9, 27, 81$ are the 4 geometric means, that we were supposed to insert between 1 and 243.

10. Now, let us understand the process of determining the sum of n terms in a geometric progression. As before, a is the first term in the series and r is the common ratio. We will use S_n to indicate the sum of n terms in the GP.

a. Case 1: Common ratio r is greater than 1

$$S_n = a + ar + ar^2 + ar^3 \ldots + ar^{n-2} + ar^{n-1}$$

Now multiply both side of the equation by r. We get

$$rS_n = ar + ar^2 + ar^3 + ar^4 \ldots + ar^{n-1} + ar^n$$

Subtracting the second equation from the first, we get

$$rS_n - S_n = ar^n - a$$

$$S_n = \frac{a(r^n - 1)}{r - 1}$$

b. Case 2: Common ratio r is less than 1 If r is less than 1, we get

$$S_n = a\frac{(1-r^n)}{1-r}$$

c. Case 3: Sum to infinite terms when common ratios, r is less than 1

When r is less than 1 and we have infinite terms in the GP, then we can see that $r^n = 0$ when *ntendstoo*∞

$$S_\infty = \frac{a}{1-r}$$

11. Let us look at an example to understand the process a bit better. If we are given the sum of an infinite number of terms in G.P is 15, and the sum of their squares is 45; and asked to find the series, we would proceed systematically as below.

Let a be the first term in the series, r be the common ratio.

$$S_\infty = \frac{a}{1-r} = 15$$

$$SSq_\infty = \frac{a^2}{1-r^2} = 45$$

Therefore,

$$\frac{SSq_\infty}{S_\infty} = \frac{\dfrac{a^2}{1-r^2}}{\dfrac{a}{1-r}} = \frac{45}{15} = 3$$

$$\frac{SSq_\infty}{S_\infty} = \frac{a}{1+r} = 3$$

Therefore, $\dfrac{\dfrac{a}{1-r}}{\dfrac{a}{1+r}} = \dfrac{15}{3} = 5$

$$\frac{1+r}{1-r} = 5 \text{ or } 6r = 4 \text{ or } r = \frac{2}{3}$$

Substituting the value of r in $\dfrac{a}{1+r} = 3$, we get $a = 5$ Thus, we can fill in the subsequent details.

12. In this example, we have simply used the formula for ∞ terms faithfully. In doing so, we have been able to arrive at the desired results systematically.

13. Let us use what we have learnt so far to determine the recurring decimal. Let p decimals be the ones that do not recurr, let us denote this sequence as P. And let q decimals repeat for ever. Let us denote this sequence by Q. Therefore the number $D = 0.P\underline{QQQQQQ}$.

$$10^p \times D = P.\underline{QQQQQQ}\ldots$$

$$10^{p+q} \times D = PQ.\underline{QQQQQQQ}\ldots$$

Subtracting the later equation from the former, we get, $[10^{p+q} - 10^p] \times D = PQ - P = P(Q-1)$ Therefore,

$$D = \frac{P(Q-1)}{10^p(10^q - 1)}.$$

The denominator has q 9s and p 0s.

14. The pattern that emerges from this example is as follows. For the numerator, take the product of the digits in the non-recurring part with the recurring part reduced by 1. For the denominator, take a number consisting of as many 9s as there are recurring figures followed by as many 0s as there are non recurring figures.

15. Let us proceed to the next level of challenges in geometric progressions. How would we handle a situation where the terms are in AP and the overall series is a GP ? Confusing ? Let us spend some time to understand the underlying concept. Let a be the first term of the series, d be the common difference and r be the common ratio. Let us now write down the sum to n terms of the desired series. This will clarify the problem statement in an instant.

$$S_n = a + (a+d)r + (a+2d)r^2 + (a+3d)r^3 + \ldots + (a+(n-1)d)r^{n-1}$$

Let us attempt to determine the sum to n terms of such a series.

Let

$$S_n = a + (a+d)r + (a+2d)r^2 + (a+3d)r^3 + \ldots + (a+(n-1)d)r^{n-1}$$

By multiplying both sides of the equation by r, we get

$$rS_n = ar + (a+d)r^2 + (a+2d)r^3 + (a+3d)r^4 + \ldots + (a+(n-1)d)r^n$$

Subtracting the later equation from the former, we get

$$(r-1)S_n = a + (d + dr + dr^2 + \ldots + dr^{n-1}) + (a+(n-1)d)r^n$$

$$S_n = \frac{a}{r-1} + \frac{dr(1-r^{n-1})}{(r-1)^2} + \frac{(a+(n-1)d)r^n}{r-1}$$

$$S_n = \frac{a}{r-1} + \frac{dr}{(r-1)^2} - \frac{dr^n}{(r-1)^2} + \frac{(a+(n-1)d)r^n}{r-1}$$

Again, if r is less than zero, r^n terms can be ignored. The sum converges to an elegant expression

$$S_n = \frac{a}{r-1} + \frac{dr}{(r-1)^2}$$

7.1 Solved problems:

1. **Sum** $\dfrac{1}{2}, \dfrac{1}{3}, \dfrac{2}{9}, \ldots$ **to 7 terms**

$$a = \frac{1}{2}, r = \frac{2}{3}, n = 7$$

$$S_n = \frac{a(1-r^n)}{1-r} = \frac{1}{2}\left[\frac{1-\left(\dfrac{2}{3}\right)^7}{1-\dfrac{2}{3}}\right]$$

$$= \frac{3}{2}\left(\frac{3^7 - 2^7}{3^7}\right) = \frac{1}{2}\left(\frac{2187 - 128}{3^6}\right)$$

$$= \frac{2059}{1458}$$

2. Sum 1, 5, 25, ……… to p terms

$$a = 1, r = 5, n = p$$

$$S_p = 1\left(\frac{5^p - 1}{5 - 4}\right) = \frac{5^p - 1}{4}$$

3. Sum 3, -4, $\dfrac{16}{3}$, …… to 2n terms.

$$a = 3, r = \frac{-4}{3} \ldots n = 2n$$

$$S_{2n} = \frac{a\left(r^{2n} - 1\right)}{r - 1} = \frac{3\left(\left(\dfrac{-4}{3}\right)^{2n} - 1\right)}{\dfrac{-4}{3} - 1}$$

$$= \frac{3\left(\left(\dfrac{16}{9}\right)^n - 1\right)}{\dfrac{-7}{3}} = \frac{-9}{7}\left(\frac{16^n}{9^n} - 1\right)$$

4. Sum $1, \sqrt{3}, 3$, …… to 12 terms.

$$a = 1, r = \sqrt{3}, n = 12$$

$$S_{12} = 1\frac{\left(\sqrt{3}^{12} - 1\right)}{\sqrt{3} - 1} = \frac{3^6 - 1}{\sqrt{3} - 1} = \frac{364(3 - 1)}{\sqrt{3} - 1}$$

$$= 304\left(\sqrt{3}+1\right)$$

5. **Sum** $\dfrac{1}{\sqrt{2}}, -2, \dfrac{8}{\sqrt{2}}, \dots$ **to 7 terms.**

$$a = 1, r = -2\sqrt{2}, n = 7$$

$$S_7 = \frac{1\left(\left(-2\sqrt{2}^{\,7}\right)-1\right)}{-2\sqrt{2}-1} = \frac{+2^7\left(\sqrt{2}\right)^7+1}{+2\sqrt{2}+1}$$

$$= \frac{1024\sqrt{2}+1}{2\sqrt{2}+1}$$

$$= \frac{\left(1024\sqrt{2}+1\right)\left(2\sqrt{2}-1\right)}{\left(2\sqrt{2}+1\right)\left(2\sqrt{2}-1\right)} = \frac{4096-1022\sqrt{2}-1}{7}$$

$$= 585-146\sqrt{2}$$

6. **Sum** $-\dfrac{1}{3}, \dfrac{1}{2}, -\dfrac{3}{4}, \dots$ **to 7 terms.**

$$a = -\frac{1}{3}, r = \frac{-3}{2}, n = 7$$

$$S_7 = \frac{\dfrac{-1}{3}\left(\left(\dfrac{-3}{2}\right)^7-1\right)}{\dfrac{-3}{2}-1} = \frac{\dfrac{-1}{3}\left(\dfrac{3^7}{2^7}+1\right)}{\dfrac{3}{2}+1}$$

$$= \frac{-1}{3}\times\frac{2}{5}\left(\frac{3^7+2^7}{2^7}\right) = \frac{-2315}{3\times5\times2^6}$$

$$= \frac{-463}{192}$$

7. Insert 3 geometric means between $2\frac{1}{4}$ and $\frac{4}{9}$

$$a = \frac{9}{4}, t_5 = \frac{4}{9}$$

$$t_5 = ar^{n-1}$$

$$\Rightarrow \frac{4}{9} = \frac{9}{4} \cdot r^4 \text{ or } r^4 = \frac{16}{81}$$

$$r = \frac{+2}{-3} = \frac{2}{3} \left| \text{Terms are } \frac{3}{2}, 1, \frac{2}{3} \right.$$

8. Insert 5 geometric means between $3\frac{5}{9}$ and $40\frac{1}{2}$

$$a = \frac{32}{9}, t_7 = \frac{81}{2}$$

$$t_7 = ar^{n-1} = \frac{32}{9} \cdot r^6 = \frac{81}{2}$$

Or $r^6 = \frac{81 \times 9}{64}$ or $r = \frac{3}{2} 3$

Means are $\frac{16}{3}, 8, 12, 18, 27$

9. Insert 6 geometric means between 14 and $-\frac{7}{64}$

$$a = 14, t_8 = \frac{7}{64}$$

$$t_8 = \frac{7}{64} = ar^7 = 14 \cdot r^7$$

$$r^7 = \frac{1}{128} \text{ or } r = \frac{1}{2}$$

Means are $7, \dfrac{7}{2}, \dfrac{7}{4}, \dfrac{7}{8}, \dfrac{7}{16}, \dfrac{7}{32}$

10. Sum the series: 1.665, -1.11, .74, ………

$$a = \dfrac{1665}{1000}, r = \dfrac{\dfrac{-111}{100}}{\dfrac{1665}{1000}} = \dfrac{-1110}{1665} = \dfrac{-2220}{3330}$$

$$= \dfrac{-2}{3}$$

$$S_\infty = \dfrac{a}{1-r} = \dfrac{\dfrac{1665}{1000}}{1 - \left(\dfrac{-2}{3}\right)} = \dfrac{3}{5} \times \dfrac{1665}{1000}$$

$$= \dfrac{999}{1000}$$

11. The sum of the first 6 terms of a G.P. is 9 times the sum of the first 3 terms, find the common ratio.

$$S_6 = 9S_3$$

Or $\dfrac{a\left(r^6 - 1\right)}{r - 1} = \dfrac{9 \cdot a\left(r^3 - 1\right)}{r - 1}$

Or $\dfrac{r^6 - 1}{r^3 - 1} = 9$ or $x^3 + 1 = 9$

$$\Rightarrow r^3 = 8 \text{ Or } r = 2$$

12. The fifth term of a G.P. is 81, and the second term is 24; find the series.

$$t_5 = 81, t_2 = 24$$

$$t_5 = ar^4 = 81; t_2 = ar = 24$$

Dividing $r^3 = \dfrac{81}{24} = \dfrac{27}{8}$

$$r = \dfrac{3}{2}$$

$$a = \dfrac{t_2}{r} = \dfrac{24}{\left(\dfrac{3}{2}\right)} = 16$$

Series $= 16, 24, 36, ...$

13. The sum of a G.P. whose common ratio is 3 is 728, and the last term is 486; find the first term.

$$r = 3,\ t_n = 486,\ S_n = 728$$

$$S_n = \dfrac{rt_n - a}{r-1} \text{ Or } 728 = \dfrac{3 \times 486 - a}{3-1}$$

Or $728 \times 2 = 3 \times 486 - a$

$$a = 1458 - 1456$$

$$= 2$$

14. Sum $1 + 2a + 3a^2 + 4a^3 + ...$ to n terms.

$$S = 1 + 2a + 3a^2 .... + na^{n-1}$$

$$aS = a + 2a^2 ..... + (n-a)a^{n-1} + na^n$$

$$S(1-a) = 1 + a + a^2 + ... + a^{n-1} - na^n$$

$$S(1-a) = \dfrac{1-a^n}{1-a} - na^n$$

$$S = \dfrac{1-a^n}{(1-a)^2} - \dfrac{na^n}{1-a}$$

15. Sum $1 + \dfrac{3}{4} + \dfrac{7}{16} + \dfrac{15}{64} + \dfrac{31}{256} + \dots$ **to infinity**

$$S = 1 + \frac{3}{4} + \frac{7}{16} + \frac{15}{64} + \frac{31}{256} + \dots$$

$$\frac{1}{4}S = \frac{1}{4} + \frac{3}{16} + \frac{7}{64} + \frac{18}{256} + \dots$$

$$\frac{3}{4}S = 1 + \frac{2}{4} + \frac{4}{16} + \frac{8}{64} + \dots$$

$$= 1 + \frac{1}{2} + \frac{1}{4} + \frac{1}{8} + \dots$$

$$= \frac{1}{1 - \frac{1}{2}} = 2$$

Or $S = \dfrac{8}{3}$

16. Sum $1 + 3x + 5x^2 + 7x^3 + 9x^4 + \dots$ **to infinity, x being < 1.**

$$S = 1 + 3x + 5x^2 + 7x^3 + \dots$$

$$xS = x + 3x^2 + 5x^3 + \dots$$

$$S(1-x) = 1 + 2x + 2x^2 + 2x^3 + \dots$$

$$= -1 + 2(1 + x + x^2 + x^3 + \dots)$$

$$= -1 + \frac{2}{1-x}$$

$$S = \frac{2}{(1-x)^2} - \frac{1}{1-x} = \frac{2-(1-x)}{(1-x)^2} = \frac{1+x}{(1-x)^2}$$

17. Sum $1 + \dfrac{3}{2} + \dfrac{5}{4} + \dfrac{7}{8} + \dots$ **to infinity.**

$$S = 1 + \frac{3}{2} + \frac{5}{4} + \frac{7}{8} + \dots$$

$$\frac{S}{2} = \frac{1}{2} + \frac{3}{4} + \frac{5}{8} + \dots$$

$$\frac{S}{2} = 1 + \frac{2}{2} + \frac{2}{4} + \frac{2}{8} + \dots$$

$$= 1 + 1 + \frac{1}{2} + \frac{1}{4} + \dots$$

$$= 3$$

Or $S = 6$

18. **Sum** $1 + 3x + 6x^2 + 10x^3 + \dots$ **to infinity, x being < 1**

$$S = 1 + 3x + 6x^2 + 10x^3 + 15x^4 + 21x^5 + \dots$$

$$xS = x + 3x^2 + 6x^2 + 10x^4 + 15x^5 + \dots$$

$$S(1-x) = 1 + 2x + 3x^2 + 4x^3 + 5x^4 + 6x^5 + \dots$$

$$xS(1-x) = x + 2x^2 + 3x^3 + 4x^4 + 5x^5 + \dots$$

$$S(1 - x - x + x^2) = 1 + x + x^2 + x^3 + \dots$$

$$S(1 - 2x + x^2) = \frac{1}{1-x}$$

$$S(1-x)^2 = \frac{1}{1-x}$$

$$S = \frac{1}{(1-x)^3}$$

19. **The sum of three numbers in G.P. is 70; if the two extremes be multiplied each by 4, and the mean by 5, the products are in A.P. find the numbers.**

$$\frac{a}{r}, a, ar$$

$$\frac{a}{r} + a + ar = 70$$

$$\frac{4a}{r}, 5a, 4ar \text{ are in AP}$$

Or $4ar - 5a = 5a - \dfrac{4a}{r}$

$$4ar + \frac{4a}{r} = 10a \Rightarrow 2ar + \frac{2a}{r} = 5a$$

Or $2r + \dfrac{2}{r} = 5$ or $2r^2 - 5r + 2 = 0$

$$(2r - 1)(r - 2) = 0$$

$$r = 2 \,\text{or}\, \frac{1}{2}$$

$$a\left(\frac{1}{r} + 1 + r\right) = 70$$

$$a = \frac{70}{2 + 1 + \dfrac{1}{2}} = \frac{140}{7} = 20$$

Numbers are $10, 20, 40$

20. **The first two terms of an infinite G. P. are together equal to 5, and every term is 3 times the sum of all the terms follow it; find the series.**

Series is $a, ar, ar^2 \ldots$

$$a + ar = 5$$

$$a = 3(ar + ar^2 + \ldots)$$

$$= 3ar(1 + r + r^2 + \ldots)$$

$$1 = \frac{3r}{1 - r} \,\text{or}\, r = \frac{1}{4}$$

$$a = \frac{5}{1+r} = 4$$

Numbers are $4, 1, \frac{1}{4}, \frac{1}{16}, \ldots$

21. Sum $x+a, x^2 +2a, x^3 +3a \ldots$ **to n terms**

$$S = x + a + x^2 + 2a + x^2 + 3a + \ldots + x^4 + na$$

$$S = \left(x + x^2 + x^3 + \ldots x^4\right) + \left(a + 2a + \ldots + na\right)$$

$$= x\left(\frac{1-x^n}{1-x}\right) + \frac{an(n+1)}{2}$$

22. If a, b, c, d be in G. P. , prove that
$$(b-c)^2 + (c-a)^2 + (d-b)^2 = (a-b)^2$$

a, b, c, d are in GP

$$b = ar, c = ar^2, d = ar^3$$

$$(b-c)^2 (c-a)^2 + (d-b)^2$$

$$= (ar - ar^2)^2 (ar^2 - a)^2 + (ar^3 - ar)^2$$

$$= a^2 r^2 (1-r)^2 + a^2 (r^2 - 1)^2 + a^2 r^2 (r^2 - 1)^2$$

$$= a^2 r^2 (1 - 2r + r^2) + a^2 (r^4 - 2r^2 + 1) + a^2 r^2 (r^4 - 2r^2 + 1)$$

$$= a^2 r^2 - 2a^2 r^2 + a^2 r^4 + a^2 r^4 - 2a^2 r^2$$
$$\qquad + a^2 + a^2 r^6 - 2a^2 r^4 + a^2 r^2$$

$$= a^2 - 2a^2 r^3 + a^2 r^6$$

$$= (a - ar^3)^2$$

$$= (a-d)^2$$

23. If the arithmetic mean between a and b is twice as great as the geometric mean, show that $a : b = 2 + \sqrt{3} : 2 - \sqrt{3}$.

$A.M = 2.GM$

$$\frac{a + b}{2} = 2\sqrt{ab}$$

$$\Rightarrow a - 4\sqrt{ab} + b = 0$$

$$\Rightarrow \left(\sqrt{a}\right)^2 - 4\sqrt{a} \cdot \sqrt{b} + \left(\sqrt{b}\right)^2 = 0$$

$$\Rightarrow \sqrt{a} = \frac{+4\sqrt{b} \pm \sqrt{\left(-4\sqrt{b}\right)^2 - 4b}}{2}$$

$$\sqrt{a} = \frac{4\sqrt{b} + \sqrt{12b}}{2} = 2\sqrt{b} + \sqrt{3}\sqrt{b}$$

$$\frac{\sqrt{a}}{\sqrt{b}} = 2 + \sqrt{3}$$

$$\frac{a}{b} = \left(2 + \sqrt{3}\right)^2 = LHS.$$

$$RHS = \frac{(2 + \sqrt{3})}{(2 - \sqrt{3})} = \frac{(2 + \sqrt{3})}{(2 - \sqrt{3})} \times \frac{(2 + \sqrt{3})}{(2 + \sqrt{3})}$$

$$= (2 + \sqrt{3})^2$$

LHS=RHS. Hence proved.

8 Harmonic Progression& related theorems

1. Definition:- Three quantities a, b, c are said to be in Harmonical Progression when $\dfrac{a}{c} = \dfrac{a-b}{b-c}$

2. Clearly, the reciprocals of the quantities in harmonical progression are in arithmetical progression.

    If a, b and c are in HP, then $\dfrac{a}{c} = \dfrac{a-b}{b-c}$.

    $\therefore a(b-c) = c(a-b)$

    Let us now divide both sides of the equation by abc

    $$\frac{1}{c} - \frac{1}{b} = \frac{1}{b} - \frac{1}{a}$$

    Thus it is clear that the reciprocal quanitities of a HP are in AP.

3. So, the basic strategy for solving problems in HP is to convert the series into an AP, solve the AP and take their reciprocals to determine the final answer. In other words, we will invert the terms and use the properties of resulting AP.

4. Let us now determine the harmonic mean between two quanities a and b

    Let a, b be the two quantities and let h be their harmonic mean.

    $\therefore \dfrac{1}{a}, \dfrac{1}{h}, \dfrac{1}{b}$ are in AP.

    $\therefore \dfrac{1}{h} - \dfrac{1}{a} = \dfrac{1}{b} - \dfrac{1}{h}$

    $\dfrac{2}{h} = \dfrac{1}{a} + \dfrac{1}{b}$

    $\therefore h = \dfrac{2ab}{a+b}$

5.   If a and b are two quantities, we know that

    a.   Arithmetic Mean $A = \dfrac{a+b}{2}$

    b.   Geometric Mean $G = \sqrt{ab}$

    c.   Harmonic Mean $H = \dfrac{2ab}{a+b}$

    d. $A \times H = \dfrac{a+b}{2} \times \dfrac{2ab}{a+b} = ab = G^2$

    e.   In other words, G is the Geometric Mean between A and H.

    f.   Let us get ready to make the next set of observations.

$$A - G = \frac{a+b}{2} - \sqrt{ab}$$

$$= \frac{a+b-2\sqrt{ab}}{2}$$

$$= \frac{\sqrt{a} - \sqrt{b}^{\,2}}{\sqrt{2}}$$

This implies that $A - G$ is always positive.

In other words, A is always greater than G.

Therefore, we can conclude that the Arithmetic Mean, Geometric Mean and Harmonic Mean are in descending order of magnitude.

6.   If the same quantity be added to, or subtracted from, all the terms of an AP, the resulting terms will form an A.P, with same common difference as before.

7.   If all the terms of an AP be multiplied or divided by the same quantity, the resulting terms will form an AP, but with a new common difference.

8. If all the terms of a GP be multiplied or divided by the same quantity, the resulting terms will form a GP. With the same common ratio as before.

9. If a, b, c, d are in GP. They are also in continued proportion.

10. Sum of first n natural numbers

    First term $= 1$

    Common difference $= 1$

    $$S_n = \frac{n}{2} \times (2 \times 1 + (n-1) \times 1)$$

    $$\therefore S_n = \frac{n(n+1)}{2}$$

11. Let us now get to the task of determining the sum of the squares of the first n natural numbers

    $$S_n = 1^2 + 2^2 + 3^2 \ldots + (n-1)^2 + n^2$$

    Now, we use a basic relationship we learnt in Elementary Algebra.

    $$n^3 - (n-1)^3 = 3n^2 - 3n + 1$$

    Similarly, $(n-1)^3 - (n-3)^3 = 3(n-1)^2 - 3(n-1) + 1$

    $\ldots$

    $\ldots$

    $$3^3 - 2^3 = 3.3^2 - 3.3 + 1$$

    $$2^3 - 1^3 = 3.2^2 - 3.2 + 1$$

    $$1^3 - 0^3 = 3.1^2 - 3.1 + 1$$

    Adding all the equations above, we have

    $$n^3 - 0^3 = 3(1^2 + 2^2 + 3^2 \ldots + n^2) - 3(1 + 2 + 3 + 4 \ldots + n) + n$$

    $$n^3 = 3S_n - 3\frac{n(n+1)}{2} + n$$

    $$3S_n = n^3 + 3\frac{n(n+1)}{2} - n$$

$$= n(n+1)(n-1) + 3\frac{n(n+1)}{2}$$

$$= \frac{n(n+1)(2n+1)}{2}$$

$$\therefore S_n = \frac{n(n+1)(2n+1}{6}$$

12. Let us now derive a formula for the sum of the cubes of the first n natural numbers.

Our strategy will be the similar to the one we used for determining the sum of squares of the first n natural numbers. Here we will start with difference of fourth powers of two successive numbers.

Let S_n represent the sum of the cubes of the first n natural numbers.

$$n^4 - (n-1)^4 = 4n^3 - 6n^2 + 4n - 1$$

$$(n-1)^4 - (n-2)^4 = 4(n-1)^3 - 6(n-1)^2 + 4(n-1) - 1$$

...

...

...

$$3^4 - 2^4 = 4.3^3 - 6.3^3 + 4.3 - 1$$

$$2^4 - 1^4 = 4.2^3 - 6.2^2 + 4.2 - 1$$

$$1^4 - 0^4 = 4.1^3 - 6.1^2 + 4.1 - 1$$

Hence, by addition,

$$n^4 = 4S_n - 6(1^2 + 2^2 + ... + n^2) + 4(1 + 2 + ... + n) - n$$

$$4S_n = n^4 + 6(1^2 + 2^2 + ... + n^2) - 4(1 + 2 + ... + n) + n$$

Substituting the formulas for sum of n natural numbers, we get

$$4S_n = n^4 + n + n(n+1)(2n+1) - 2n(n+1)$$

$$= n(n+1)(n^2 - n + 1 + 2n + 1 - 2)$$

$$= n(n+1)(n^2 + n)$$

$$\therefore S_n = \left(\frac{n(n+1)^2}{2} \right)$$

13. If we can visualize the problem in the form of a picture, or draw a diagram to represent our understanding of the problem, solutions will pop out of the diagram. Clarity in thought and representation of our understanding is key to successful problem solving. Let us consider a couple of examples to demonstrate the same.

14. Let us start off with an arrangement of balls in the form of a pyramid on a square base. The problem is to determine the total number of balls in the pyramid.

   a. The base of the pyramid has $n \times n = n^2$ balls. The layer above it, has $(n-1) \times (n-1) = (n-1)^2$ balls and so on. We have one ball at the top of the pyramid.

   b. The total number of balls
       $$= n^2 + (n-1)^2 + (n-3)^2 \ldots + 3^2 + 2^2 + 1^2$$

   c. The total number of balls is simply the sum of the squares of the n natural number. Therefore, the sum of balls is $\frac{n(n+1)(2n+1)}{6}$.

15. What would the solution look like if the base of the pyramid was an equilateral triangle instead of a square? No issues.

   a. Let there be n balls in one side of the equilateral triangle at the base of the pyramid.

   b. The total number of balls in the base of the pyramid is $\frac{n(n+1)}{2}$

   c. The total number of balls in the pyramid is
       $$S_n = \frac{1}{2} = \frac{1}{2} \left(\sum n^2 + \sum n \right)$$

d. We can substitute the required formulas and determine the final sum of balls $= \dfrac{n(n+1)(n+2)}{6}$

16. Like we said in the opening section of the book. Never panic while solving problems. There is a representation of every problem - a formulation that makes things simple for us.

8.1 Solved problems

1. Find the fourth term in each of the following series:

a. $2, 2\dfrac{1}{2}, 3\dfrac{1}{3}, \ldots$

These are in HP

Corresponding AP $= \dfrac{1}{2}, \dfrac{2}{5}, \dfrac{3}{10} = \dfrac{5}{10}, \dfrac{4}{10}, \dfrac{3}{10}$

Next term is $\dfrac{2}{10} = \dfrac{1}{5}$

Next term $= 5$

b. $2, 2\dfrac{1}{2}, 3, \ldots$

These are in AP with a common difference of $\dfrac{1}{2}$

Next term $= 3\dfrac{1}{2} = \left[3 + \left(3 - \dfrac{5}{2}\right)\right]$

c. $2, 2\dfrac{1}{2}, 3\dfrac{1}{8}, \ldots$

These are in GP with a common ratio of $\dfrac{5}{4}$

$$a = 2, r = \frac{5}{4}$$

$$\text{Next term} = \frac{25}{8} \times \frac{5}{4} = \frac{125}{32}$$

2. Insert two harmonic means between 5 and 11.

$$\frac{1}{11}, -, -, \frac{1}{5} \text{ are in AP}$$

$$a = \frac{1}{11}, a + 3d = \frac{1}{5}$$

$$d = \frac{\frac{1}{5} - \frac{1}{11}}{3} = \frac{6}{5 \times 11 \times 3} = \frac{2}{55}$$

Numbers of AP are $\frac{1}{11} + \frac{2}{55} = \frac{7}{55}$

And $\frac{7}{55} + \frac{2}{55} = \frac{9}{55}$

HP numbers are $\frac{55}{9}$ and $\frac{55}{7}$

3. Insert four harmonic means between $\frac{2}{3}$ and $\frac{2}{13}$.

$$\text{AP } t_1 = \frac{3}{2}, t_6 = \frac{13}{2} = \frac{3}{2} + 5d$$

$$d = \frac{\frac{13}{2} - \frac{3}{2}}{5} = 1$$

AP terms are, $\frac{3}{2}, \frac{5}{2}, \frac{7}{2}$ and $\frac{9}{2}, \frac{11}{2}$.

HP means are $\frac{2}{5}, \frac{2}{7}, \frac{2}{9}, \frac{2}{11}$

4. **If 12 and $9\dfrac{3}{5}$ are the geometric and harmonic means, respectively, between two numbers, find them.**

    Numbers be a and b

    $$\text{GM}=12=\sqrt{ab}\Rightarrow ab=12^2=144$$

    $$\text{HM}=\frac{48}{5}\frac{2ab}{a+b}$$

    Or $a+b=\dfrac{5}{24}\times ab=\dfrac{5}{24}\times144=30$

    $$a=30-b$$

    $$ab=b(30-b)=144$$

    Or $b^2-30b+144=0$

    $$(b-6)(b-24)=0$$

    $$b=6\,\text{or}\,24$$

    $$a=24\,\text{or}\,6$$

5. **If the harmonic mean between two quantities is as to their geometric means as 12 to 13, prove that the quantities are in the ratio of 4 to 9.**

    $$\frac{\text{HM}}{\text{GM}}=\frac{12}{13}\Rightarrow\frac{\dfrac{2ab}{a+b}}{\sqrt{ab}}=\frac{12}{13}$$

    $$\frac{2\sqrt{ab}}{a+b}=\frac{12}{13}\Rightarrow\frac{\cancel{4}\,ab}{(a+b)^2}=\frac{\cancel{144}^{\,36}}{169}$$

    $$\frac{ab}{(a+b)^2}=\frac{36}{169}\Rightarrow\frac{\dfrac{9}{b}}{\left(\dfrac{9}{b}+1\right)}=\frac{36}{169}$$

$$36(r^2 + 2r + 1) = 169r \left| r = \frac{9}{b} \right.$$

$$36r^2 - 97r + 36 = 0 \Rightarrow (4r - 9)(9r - 4) = 0$$

$$r = \frac{9}{4} \text{ or } \frac{4}{9}$$

6. **If a, b, c, be in H. P. show that** $a : a - b = a + c : a - c$.

$b = $ HM of a and c

$$b = \frac{2ac}{a+c}$$

$$\frac{a}{a-b} = \frac{a}{a - \dfrac{2ac}{a+c}} = \frac{\cancel{a}(a+c)}{\cancel{a}(a+c) - 2\cancel{a}c} = \frac{a+c}{a-c}$$

7. **If the mthterm of a H.P. be equal to n, and the nth term be equal to m prove that the** $(m+n)$ **th term is equal to** $\dfrac{mn}{m+n}$

m^{th} term of a HP = n

$$t_m = \frac{1}{n} = a + (m-1)d$$

n^{th} term $= m$

$$t_n = \frac{1}{m} = a + (n-1)d$$

Subtracting:

$$\frac{1}{n} - \frac{1}{m} = (m-n)d$$

$$\frac{m-n}{mn} = (m-n)d$$

$$d = \frac{1}{mn}$$

$$a = \frac{1}{n} - (m-1)d = \frac{1}{n} - \frac{m-1}{mn} = \frac{1}{mn}$$

t_{m+n} of the AP $= a + (m+n-1)d$

$$= \frac{1}{mn} + \frac{(m+n-1)}{mn} = \frac{m+n}{mn}$$

HP term is $\dfrac{mn}{m+n}$

8. **If the pth, qth, rthterms of a H.P. be a, b, c respectively, prove that** $(q-r)bc + (r-p)ca + (p-q)ab = 0$

For the AP

$$t_p = \frac{1}{a}, t_q = \frac{1}{b}; t_r = \frac{1}{c}$$

$$\frac{1}{a} = t_1 + (p-1)d; \frac{1}{b} = t_1 + (q-1)d; \frac{1}{c} = t_1 + (r-1)d$$

$$\frac{1}{a} - \frac{1}{b} = (p-q)d = \frac{(b-a)}{ab}$$

Or $(p-q) \times ab = \dfrac{(b-a)}{d}$

Similarly: $\dfrac{1}{b} - \dfrac{1}{c} = (q-r)d = \dfrac{c-b}{bc}$

Or $(q-r)bc = \dfrac{c-b}{d}$

$$\frac{1}{c} - \frac{1}{a} = (r-p)d = \frac{a-c}{ac}$$

$(r-p)ac = \dfrac{a-c}{d}$

68

$$(a-r)bc+(r-p)ca+(p-q)ab$$

$$=\frac{c-b}{d}+\frac{a-c}{d}+\frac{b-a}{d}$$

$$=0$$

9. **If the b is the harmonic mean between a and c, prove that**
$$\frac{1}{b-a}+\frac{1}{b-c}=\frac{1}{a}+\frac{1}{c}.$$

$b = $ HM of a and c

$$b=\frac{2ac}{a+c}$$

$$\frac{1}{b-a}+\frac{1}{b-c}=\frac{1}{\dfrac{2ac}{a+c}-a}+\frac{1}{\dfrac{2ac}{a+c}-c}$$

$$=\frac{a+c}{2ac-a^2-ac}+\frac{a+c}{2ac-c^2-ac}$$

$$=\frac{a+c}{a(c-a)}+\frac{a+c}{c(a-c)}$$

$$=\frac{c(a+c)}{ac(c-a)}-\frac{a(a+c)}{ac(c-a)}$$

$$=\frac{ac+c^2-a^2-ac}{ac(c-a)}=\frac{c^2-a^2}{ac(c-a)}=\frac{c+a}{ac}$$

$$=\frac{1}{a}+\frac{1}{c}$$

10. **Find the sum of n terms of a series whose nth term is** $3n^2-n$

S_n if $t_n=3n^2-n$

$$\sum(3n^2-n)=3\sum n^2-\sum n$$

$$= \frac{3n(n+1)(2n+1)}{6} - \frac{n(n+1)}{2}$$

$$= \frac{n(n+1)(2n+1) - n(n+1)}{2} = \frac{n(n+1)(2n+1-1)}{2}$$

$$= \frac{2n^2(n+1)}{2}$$

$$= n^2(n+1)$$

Find the number of shot in:

11. A square pile, having 15 shot in each side of the base.

Square pile, $n = 15$ on the base

$$S = \overline{z}n^2 = \frac{n(n+1)2n+1)}{6}$$

$$= \frac{15 \times 16 \times 31}{6} = 1240$$

12. A triangular pile, having 18 shot in each side of the base.

Triangular base

$$S_n = \frac{1}{2}\left(\Sigma n^2 + \Sigma n\right) = \frac{n(n+1)(n+2)}{6}$$

$$n = 18 \Rightarrow S_n = \frac{18 \times 19 \times 20}{6} = 1140$$

13. A rectangular pile, the length and the breadth of the base containing 50 and 28 shot respectively.

Rectangular

$$S_{m,n} = \frac{n(n+1)(3m-n+1)}{6}$$

$$S_{50,26} = \frac{28 \times 29 \times (3 \times 50 - 26 + 1)}{6}$$

$$= 2470$$

14. An incomplete triangular pile, a side of the base having 25 shot and a side of the top 14.

Incomplete triangle, base 25, top 14

Full triangle 25 – Full triangle13

$$= S_{25} - S_{13} = \frac{25 \times (25+1) \times (25+2)}{6} - \frac{13 \times (13+1) \times (13+2)}{6}$$

$$= 2470$$

15. An incomplete square pile of 27 courses, having 40 shot in each side of the base.

Incomplete square pile with a base 40 shots on each side; the layer above it will have 39 shots on each side, then 38 shots on each side and so on. We have 27 such layers.

$$= 40^2 + 39^2 + \ldots + (40 - 27 + 1)^2$$

= {Sum of squares from 40 to 1} MINUS {Sum of squares from 13 to 1}

$$= 40^2 + 39^2 + \ldots 1^2 - \frac{13 \times 14 \times (2 \times 13 + 1)}{6}$$

$$= \sum_{n=0\to40} n^2 - \sum_{n=0-13} n^2 = 21431$$

$$= 21431$$

16. The number of shots in a complete rectangular pile is 24395; if there are 34 shot in the bredth of the base, how many are there in its length?

We are give a rectangular pile with the following information.

$$S_{m,n} = 24395; n = 34$$

$$\Rightarrow \frac{n(n+1)(3m-n+1)}{6} = S_{m,n} = 24395$$

$$\Rightarrow \frac{34 \times 35}{6}(3m-34-1) = S_{m,n} = 24395$$

$$3m-33 = 123 \mid 3m = 156$$

$$m = 52$$

17. The number of shot in the top layer of a square pile is 169, and in the lowest layer is 1089; how many shot does the pile contain?

Lowest layer has 1089

Shot in the side $= \sqrt{1089} = 33$

Top layer has 169

Shot per side $= \sqrt{169} = 13$

Num shot per side $= S_{33} - S_{12} \mid S_n = \frac{n(n+1)(2n+1)}{6}$

$$= \frac{33 \times 34 \times (33 \times 2 + 1)}{6} - 12 \times 13 \times \frac{(12 \times 2 + 1)}{6}$$

$$= 11879$$

18. Find the number of shot in an incomplete rectangular pile the number of shot in the sides of its upper of its upper course being 11 and 18 and the number in the shorter of its lowest course being 30.

Lower has $m_1, n_1; n_1 = 30$

Upper has $m_2, n_2; n_2 = 11; m_2 = 18$

In rest pile Each layer has m, n less by 1from the previous layer. So change in m = change m n

$$\Rightarrow m_1 = n_1 + m_2 - n_2 = 37$$

$$\text{Total} = S_{37,30} - S_{17,10} = 11940$$

$$= \frac{30 \times 31 \times (3 \times 37 - 30 + 1)}{6} - \frac{10 \times 11 \times (3 \times 17 - 10 + 1)}{6}$$

19. **What is the number of shot required to complete a rectangular pile having 15 and 6 shot in the longer and shorter side, respectively of its upper course?**

If upper courses has 15, 6

Next course has 14, 5; next is 13, 4 and so on

$$\text{Number to complete} = S_{14,5} = \frac{5 \times 6 \times (3 \times 14 - 5 + 1)}{6}$$

$$= 190$$

20. **The number of shot in a triangular pile is greater by 150 than half the number of shot in a square pile, the number of layers in each being the same; find the number of shot in the lower layer of the triangular pile**

Let triangle be t upon a side, square be n

If number of layers is the same, then $t = n$

Hence $S_{t,n} - \dfrac{1}{2} S_{s,n} = 150$

$$\frac{n(n+1)(n+2)}{6} - \frac{1}{2}\frac{n(n+1)(2n+1)}{6} = 150$$

$$\frac{n(n+1)}{6}\left[\frac{n+2}{1} - \frac{2n+1}{2}\right] = 150$$

$$3 \times \frac{n(n+1)}{12} = 150 \Rightarrow n(n+1) = 600$$

$$n = 24$$

9 Scales of Notation

1. Let us commence our discussions by considering our decimal number system. The numbers are formed by symbols $0,1,2,3,4,5,6,7,8$ and 9.

2. The numbers $0,1,2,3,4,5,6,7,8$ and 9 form the digits in the decimal number system.

   The decimal number system is on a scale whose radix is 10.

3. In general, if in the scale whose radix is r we denote the digits, beginning with that in the unit place, by $a_0, a_1, a_2 ... a_n$; then the number so formed will be represented by $a_n r^n + a_{n-1} r^{n-1} + a_{n-2} r^{n-2} + ... + a_2 r^2 + a_1 r + a_0$, where the co-efficient $a_n, a_{n-1}, ... a_0$ are integers, all less than r, of which any one or more after the first may be zero.

   Hence in this scale the digits are r in number, their values ranging from 0 to $r-1$.

4. The ordinary numbers with which we are acquainted in Arithmetic are expressed by means of multiples of powers of 10; for instance

   $$25 = 2 \times 10 + 5$$

   $$4705 = 4 \times 10^3 + 7 \times 10^2 + 0 \times 10 + 5$$

This method of representing numbers is called the common or denary scale of notation, and ten is said to be the radix of the scale.

In like manner any number other than 10 may be taken as the radix of a scale of notation.

5. The names Binary, Ternary, Quaternary, Quinary, Senary, Septenary, Octenary, Nonary, Denary, Undenary and Duodenary are used to denote the scales corresponding to the values two, three . twelve of the radix.

In the Undenary, duodenary scales we shall require symbols to represent the digits which are greater than nine. In such cases we use the letters $A, B...$ the numbers following 9.

6.   It is especially worthy of notice that in every scale 10 is the symbol not for `ten' but for the radix itself.

For example, in the binary system `10' stands for 2 and not 10. Similarly `10' in octal stands for eight and `10' in hexadecimal system stands for sixteen.

7.   The ordinary operations of Arithmetic may be performed in any scale; but, bearing in mind that the successive powers of the radix are no longer powers of ten, in determining the carrying figures we must not divide by ten, by the radix of the scale in question.

For example, if we wanted to subtract 371532 from 530225, and multiply the difference by 27, in a scale of eight, the octal system---, we will go through the following steps.

Step 1:

$$
\begin{array}{r}
530225 \\
371532 \\
\hline
136473
\end{array}
$$

Step 2:

$$
\begin{array}{r}
136473 \\
\times 27 \\
\hline
1226235 \\
275166 \\
\hline
4200115.
\end{array}
$$

Let us understand the numerical manipulation in this case. After the first figure of the subtraction, since we cannot take 3 form 2 we add 8; thus we have to take 3 from ten, which leaves 7; then 6 from ten, which leaves 4; then 2 from eight which leaves 6; and soon.

Again, in multiplying by 7, we have $3 \times 7 = twentyone = 2 \times 8 + 5$; We therefore, put down 5 and carry 2.

Next $7 \times 7 + 2 = fiftyone = 6 \times 8 + 3$.

Put down 3 and carry 6; and soon, until the multiplication is completed.

In the addition, $3+6 = nine = 1 \times 8 + 1$. We therefore put down 1 and carry 1.

Similarly $2+6+1 = nine = 1 \times 8 + 1$;

And $6+1+1 = eight = 1 \times 8 + 0$;

Now, we will learn to express a given integral number in any proposed scale. Let N be the given number, and r the radix of the proposed scale.

Let $a_0, a_1, a_2 \dots a_n$ be the required digits by which N is to be expressed, beginning with that in the unit's place, then

$$N = a_n r^n + a_{n-1} r^{n-1} + \dots + a_2 r^2 + a_1 r^1 + a_0.$$

We have now to find the values of $a_0, a_1, a_2 \dots a_n$.

Divide N by r, then the remainder is a_0, and the quotient is

$$a_n r^{n-1} + a_{n-1} r^{n-2} + \dots + a_2 r + a_1.$$

If this quotient is divided by r, the remainder is a_1;

If the next quotient is divided by r, the reminder is a_2; and so on, until there is no further quotient.

Thus all the required digits $a_0, a_1, a_2 \dots a_n$ are determined by successive divisions by the radix of the proposed scale.

8. Fractions may also be expressed in any scale notation.

(a) $0 \cdot 25$ in scale ten denote $\dfrac{2}{10} + \dfrac{5}{10^2}$;

(b) $0 \cdot 25$ in scale six denotes $\dfrac{2}{6} + \dfrac{5}{6^2}$;

(c) $0 \cdot 25$ in scale r denotes $\dfrac{2}{r} + \dfrac{5}{r^2}$.

9. Fractions thus expressed in an form analogous to that of ordinary decimal fractions are called radix fractions, and the point is called the radix point.

10. Let us now look how to express a given radix fraction in any proposed scale.

Let F be a fraction, then

$$F = \frac{b_1}{r} + \frac{b_2}{r^2} + \frac{b_3}{r^3} + \ldots$$

We have now to find the values of $b_1, b_2, b_3 \ldots$ multiply both sides of the equations by r, then

$$rF = b_1 + \frac{b_2}{r} + \frac{b_3}{r^2} + \ldots;$$

Hence b_1 is equal r; then, as before, b_2 is the integral part of rF_1; and similarly by successive multiplications by r, each of the digits may be found, and the fraction expressed in the proposed scale.

(a) In this process if one of the product is an integer, the process terminates.

(b) If none of the products is an integer the process will never terminate.

11. In any scale of notation of which the radix r, the sum of the digits of any whole number divided by $r-1$ will leave the same remainder as the whole number divided by $r-1$.

Let N denotes the number, $a_0, a_1, a_2 \ldots a_n$ the digits beginning with that in the units place, and S the sum of the digits, then

$$N = a_0 + a_1 r + a_2 r^2 + \ldots + a_{n-1} r^{n-1} + a_n r^n;$$

$$S = a_0 + a_1 + a_2 + \ldots + a_n$$

$$\ldots N - S = a_1(r-1) + a_2(r^2 - 1) + \ldots + a_{n-1}(r^{n-1} - 1) + a_n(r^n - 1).$$

Now every term on the right hand side is divided by $r-1$

$$\ldots \quad \frac{N-S}{r-1} = \text{an integer.}$$

i.e. $\dfrac{N}{r-1} = I + \dfrac{S}{r-1}$,

Hence a number in scale r will be divisible by $r-1$ when the sum of its digits is divisible by $r-1$.

By taking r = 10 we learn from the above proposition that a number divided by 9 will leave the same remainder as the sum of the digits divided by 9. The rule is known as ``Casting out the nines".

12. Question : Can the product of 31256 and 8427 be 263395312?

Answer: The sum of the digits of the multiplicand, multiplier, and product are 17, and 34 respectively; again, the sums of the digits of these three numbers are 8, 3 and 7.

$8 \times 3 = 24$, which has 6 for the sum of the digits; thus we have two different remainder 6 and 7, and the multiplication is incorrect.

13. In N denote any number in the scale of r, and D denote the difference, supposed positive, between the sums of the digits in the odd and the even places; then N -- D or N + D is a multiple of r + 1.

Let $a_0, a_1, a_2 \ldots a_n$ denote the digits beginning with that in the unit's place; then

$$N = a_0 + a_1 r + a_2 r^2 + \ldots + a_{n-1} r^{n-1} + a_n r^n$$
$$N - a_0 + a_1 - a_2 + a_3 - \ldots$$
$$= a_1(r+1) + a_2(r^2 - 1) + a_3(r^3 + 1) + \ldots ;$$

And the last term on the right will be $a_n(r^n + 1)$ or $a_n(r^n - 1)$ according as n is odd or even. Thus every term on the right is divisible by r + 1.

$$\text{Hence} \quad \frac{N - (a_0 - a_1 + a_2 - a_3 + \ldots)}{r+1} = \text{an integer.}$$

Now $a_0 - a_1 + a_2 - a_3 + ... = \pm D$

$\dfrac{N \mp D}{r+1}$ is an integer.

Which proves the proportion.

14. If the sum of the digits in the even places is equal to the sum of the digits in the odd places, D = 0, and N is divisible by r + 1.

9.1 Solved problems

1. **Add together 23241, 4032, 300421 in the scale of five.**

$$
\begin{array}{r}
23241 \\
4032 \\
300421 \\
\hline
333244
\end{array}
\quad \text{In base 5}
\left|
\begin{array}{l}
9 = 5 + 4 \\
7 = 5 + 2
\end{array}
\right.
$$

2. **Find the sum of the nonary numbers 303478, 150732, 264305.**

Nonary = base9

$$
\begin{array}{r}
303478 \\
150732 \\
264305 \\
\hline
728626
\end{array}
\left|
\begin{array}{l}
15 = 9 + 6 \\
11 = 9 + 2
\end{array}
\right.
$$

3. **Subtract 1732765 from 3673124 in the scale of eight.**

$$
\begin{array}{r}
3673124 \\
1732765 \\
\hline
1740137
\end{array}
$$

Base 8

4. From 3et756 take 2e46t2 in the duodenary scale.

Base 12,

$$3et756$$
$$2e46t2$$
$$\overline{106074}$$

5. Divide the difference between 1131315 and 235143 by 4 in the scale of six.

Base 61131315

$$235143 \quad 4\sqrt{452132}\,(112022$$

$$\frac{4}{5}$$

$$\frac{4}{12}$$

$$\frac{12}{13}$$

$$\frac{12}{12}$$

$$= 112022$$

6. Multiply 6431 by 35 in the scale of seven.

Base 7

$$35 \times 6431$$
$$\overline{45115}$$
$$25623$$
$$\overline{334345}$$

7. Find the product of the nonary numbers 4685, 3483

Base 9

$$\frac{4685 \times 3483}{18656}$$

$$31336$$

$$23250$$

$$\frac{15163}{17832126}$$

8. Divide 102432 by 36 in the scale of seven.

Divide 102432 by in base 7

$$36)\overline{\smash{\big)}\,102432}\,(1625$$
$$\underline{36}$$
$$334$$
$$\underline{321}$$
$$133$$
$$\underline{105}$$
$$252$$
$$\underline{252}$$
$$0$$

Ans $= 1625$

9. In the ternary scale subtract 121012 from 11022201, and divide the result by 1201.

In base 3,

$$11022201$$
$$\underline{-121012}$$
$$10201112$$

$$1201)\dfrac{\sqrt{10201112}}{10102}(2012$$

$$2211$$

$$\underline{1201}$$

$$10102$$

$$\underline{10102}$$

Answer $= 2010$

10. Express 4954 in the scale of seven.

4954 in base 7

7)4954

7)707....5

7)101.....0

7)14....3

2......0

Ans $= 20305$

11. Express 624 in the scale of five.

624 in base 5

5)624

124.....4 $\bigg|\dfrac{620}{5} = 124$

24.....4

4.....4

$= 4444$

12. Express 206 in the binary scale.

206 in base 2

2)206

103....0

51....1

25....1

12....1

6....0

3....0

1....1

$= 11001110$

13. Express 1458 in the scale of three.

1458 in base 3

$$\frac{1458}{3}$$

$= 1286$

3)1458

486.....0

162.....0

54.....0

18.....0

6.....0

2.....0

$= 2000000$

14. Express 5381 in powers of nine

5381 in base 9

$$9)5381$$
$$597.....8$$
$$66.....3$$
$$7.....3$$
$$= 7338$$
$$= 7 \times 9^3 + 3 \times 9^2 + 3 \times 9^1 + 8$$

15. Transform 212231 from scale four to scale five.

Transform 212231 bases 4 to base 5

$$212231 = 2 \times 4^5 + 1 \times 4^4 + 2 \times 4^3 + 2 \times 4^2 + 3 \times 4^1 + 1$$
$$= 2 \times 1059 + 256 + 2 \times 64 + 2 \times 16 + 3 \times 4 + 1$$
$$= 2477 \text{ in base } 10$$

In base 5,

$$5)2477$$
$$495.....2$$
$$99.....0$$
$$19.....4$$
$$3.....4$$
$$= 34402$$

Express the duodenary number $398e$ in powers of 10.

Convert 398e into base 10 from base 12

$$= 3 \times 12^3 + 9 \times 12^2 + 8 \times 12 + 11$$
$$= 3 \times 1728 + 9 \times 144 + 8 \times 12 + 11$$
$$= 6587$$

16. Transform $6t12$ from scale twelve to scale eleven.

Transform 6t12 from base 12 to base 11

e)6t12

$756.....8$

$\quad 81.....7 \quad$ 6t12 $= 11822$

$\qquad 8.....9$

$= 8978$

e)11822

$\quad 1074....8$

$\qquad 97....7$

$\qquad 8.....9$

$= 8978$

17. Transform 23861 from scale nine to scale eight.

Transform 23861 from base 9 to base 8

$$23861_9 = 2 \times 9^4 + 3 \times 9^3 + 8 \times 9^2 + 6 \times 9 + 1$$
$$= 16012$$

In base 8,

$\quad$ 8)16012

$\qquad 2001.....4$

$\qquad 250......1$

$\qquad 31......2$

$\qquad 3......7$

$= 37214_8$

18. Prove that 1.331 is a perfect cube in any scale whose radix is greater than three.

1.331 in base r, $r > 3$ since digits up to 3

$$= 1 + \frac{3}{r} + \frac{3}{r^2} + \frac{1}{r^3}$$

$$= \frac{1}{r^3}\left(r^3 + 3r^2 + 3r + 1\right)$$

$$= \frac{1}{r^3}(r+1)^3$$

$$= \left(\frac{r+1}{r}\right)^3 = \left(1+\frac{1}{r}\right)^3$$

$= (1.1)_r^3$ Is a perfect cube

10 Surds and imaginary quantities

1. The process of rationalizing an expression is to attempt to convert the denominator of an expression into a form that is devoid of fractional powers. Often we come across expressions which contain square roots or cube roots of variables. Rationalizing the expression will involve mathematical manipulations for eliminating these terms by multiplying and dividing the expression with other expressions without changing the value of the original expression.

2. We know that any expression of the form $\dfrac{a}{\sqrt{b}+\sqrt{c}}$ can be rationalized by multiplying the numerator and denominator by $\sqrt{b}-\sqrt{c}$, the conjugate to the denominator.

How does this work ?

The conjugate of the denominator is $\sqrt{b}-\sqrt{c}$. We multiply and divide the original expression by the conjugate $\sqrt{b}-\sqrt{c}$.

$$\frac{a}{\sqrt{b}+\sqrt{c}}$$

$$= \frac{a}{\sqrt{b}+\sqrt{c}} \times \frac{\sqrt{b}-\sqrt{c}}{\sqrt{b}-\sqrt{c}}$$

$$= \frac{a \times (\sqrt{b}-\sqrt{c})}{\sqrt{b}^2 - \sqrt{c}^2}$$

$$= \frac{a \times (\sqrt{b}-\sqrt{c})}{b-c}$$

$\therefore$, the expression is said to be rationalized.

3. Similarly, in the case of fraction of the form $\dfrac{a}{\sqrt{b}+\sqrt{c}+\sqrt{d}}$ where the denominator involves three quadratic surds.

4. Here we will require two steps to rationalize the denominator. First, multiply both numerator and denominator by

$$\sqrt{b}+\sqrt{c}-\sqrt{d}$$

The denominator becomes $(\sqrt{b}+\sqrt{c})^2-(\sqrt{d})^2$ or

$$b+c-d+2\sqrt{bc}\ .$$

We then multiply both numerator and denominator by $(b+c-d)-2\sqrt{bc}$, the denominator becomes $(b+c-d)^2-4bc$, which is a rational quantity.

Mathematically, this can be expressed by the following steps. We hope that this will help in making things clear and unambiguous.

$$\frac{a}{\sqrt{b}+\sqrt{c}+\sqrt{d}}$$

$$=\frac{a}{\sqrt{b}+\sqrt{c}+\sqrt{d}}\times\frac{\sqrt{b}+\sqrt{c}-\sqrt{d}}{\sqrt{b}+\sqrt{c}-\sqrt{d}}$$

$$=\frac{a(\sqrt{b}+\sqrt{c}-\sqrt{d})}{b+c-d+2\sqrt{bc}}$$

$$=\frac{a(\sqrt{b}+\sqrt{c}-\sqrt{d})}{b+c-d+2\sqrt{bc}}\times\frac{(b+c-d)-2\sqrt{bc}}{(b+c-d)-2\sqrt{bc}}$$

Now the final denominator becomes $(b+c-d)^2-4bc$

5. Let us now work on determining the factors which will rationalize a binomial surd. Two cases arise. We will look at each of these cases systematically.

Case 1: Suppose the given surd is $\sqrt[p]{a} - \sqrt[q]{b}$

Let $x = \sqrt[p]{a}; y = \sqrt[q]{b}$. Let n be the LCM of p and q. This implies that x^n and y^n are both rational.

Now, we know that $x^n - y^n$ is divisible by $x - y$ for all values of n, and $x^n - y^n = (x-y)(x^{n-1} + x^{n-2}y + x^{n-3}y^2 + \ldots + y^{n-1})$

Thus the rationalizing factor is $x^{n-1} + x^{n-2}y + x^{n-3}y^2 + \ldots + y^{n-1}$.

And the resultant product is $x^n - y^n$.

Case 2: Suppose the given surd is $\sqrt[p]{a} + \sqrt[q]{b}$

Let x, y, n have the same meaning as before, then

Case a: If n is even, $x^n - y^n$ is divisible by $x + y$

$x^n - y^n = (x+y)(x^{n-1} - x^{n-2}y + \ldots + xy^{n-2} - y^{n-1})$

Thus the rationalizing factor is $x^{n-1} - x^{n-2}y + \ldots + xy^{n-2} - y^{n-1}$.

And the resultant product is $x^n - y^n$

Case b: If n is odd, $x^n + y^n$ is divisible by $x + y$

$x^n + y^n = (x+y)(x^{n-1} - x^{n-2}y + \ldots - xy^{n-2} + y^{n-1})$

Thus the rationalizing factor is $x^{n-1} - x^{n-2}y + \ldots - xy^{n-2} + y^{n-1}$.

The resultant product is $x^n + y^n$.

6. If $\sqrt[3]{a+\sqrt{b}} = x+\sqrt{y}$ then $\sqrt[3]{a-\sqrt{b}} = x-\sqrt{y}$

   $$\sqrt[3]{a+\sqrt{b}} = x+\sqrt{y}$$

   Cubing both sides of the expression, we get

   $$(\sqrt[3]{a+\sqrt{b}})^3 = (x+\sqrt{y})^3$$

   $$a+\sqrt{b} = x^3 + 3x^2\sqrt{y} + 3xy + y\sqrt{y}$$

   Equating rational and irrational parts, we get

   $$a = x^3 + 3xy, \sqrt{b} = 3x^2\sqrt{y} + y\sqrt{y}$$

   $$a-\sqrt{b} = x^3 + 3xy - (3x^2\sqrt{y} + y\sqrt{y}) = (x-\sqrt{y})^3$$

   $$\sqrt[3]{a-\sqrt{b}} = x-\sqrt{y}$$

7. While we deal with the concepts related to binomial theorem, you will notice that:

   If $\sqrt[n]{a+\sqrt{b}} = x+\sqrt{y}$, then $\sqrt[n]{a-\sqrt{b}} = x-\sqrt{y}$

We will not dive deep into this generalization at this time. We just wanted you start looking for these patterns in Algebra as you work through these concepts.

Imaginary Quantities

1. When the quantity under the radical sign is negative, we can no longer consider the symbol $\sqrt{}$ as indicating a possible arithmetical operation.

2. Now, $\sqrt{a}$ may be defined as a symbol which obeys the relation $\sqrt{a} \times \sqrt{a} = a$

3. How do we handle $\sqrt{-a}$?

(a) The $\sqrt{-a}$ is simply some quantity which obeys the equation $\sqrt{-a} \times \sqrt{-a} = -a$.

(b) Therefore, by definition $\sqrt{-1} \times \sqrt{-1} = -1$

(c) $(\sqrt{a} \times \sqrt{-1}) \times (\sqrt{a} \times \sqrt{-1}) = a \times -1$

(d) Therefore, it is clear that $(\sqrt{a}\sqrt{-1})^2 = -a$

(e) Thus the product $\sqrt{a} \times \sqrt{-1}$ may be regarded as equivalent to the imaginary quality $\sqrt{-a}$

(f) So, every occassion where we deal with a square root of a negative quanitity may be seen as a square root of magnitude of the quanity multiplied by square root of minus 1. In other words, $\sqrt{-a} = \sqrt{a}\sqrt{-1}$.

(g) We represent the imaginary quanity $sqrt-1$ by the symbol i. Therefore $\sqrt{-a} = \sqrt{a}i$.

(h) This implies that i is an imaginary quantity which obeys the equation $i^2 = -1$.

4. The expression $a + ib$ is the general representation of all expressions, where $i = \sqrt{-1}$.

5. a and b are real quanities but are not necessarily rational. $a + b\sqrt{-1}$ may be taken as the general type of all imaginary expressions. (Here a and b are real quantities but not necessarily rational.)

6. In dealing with imaginary quantities, we can apply the following laws.

   (a) If $(a + ib) = (c + id), \Rightarrow a = c; b = d$. If the corresponding real parts and imaginary parts of two imaginary quanitities are equal, then the imaginary numbers are equal.

   (b) $(a + ib) \pm (c + id) = (a \pm c) + i(b \pm d)$. In other words, if we want to add or subtract imaginary expressions, we add the real and imaginary part seperately and put them together in the form of an imaginary expression.

(c) $(a+ib)\times(c+id)$

$= ac+iad+ibc+i^2bd$, we know that $i^2 = -1$,

$= (ac-bd)+i(ad+bc)$

(d) $\dfrac{a+ib}{c+id}$

$= \dfrac{a+ib}{c+id}\times\dfrac{c-id}{c-id}$

$= \dfrac{(a+ib)(c-id)}{c^2+d^2}$

$= \dfrac{ac+bd+i(bc-ad)}{c^2+d^2}$

(e) It can be concluded that the sum, difference, product and quotient of two imaginary expression is another imaginary expression of the same form.

7. Let us quickly look at the expression $a+ib$ and a couple of observations.

(a) If $a=0$, we are left with a purely imaginary number.

(b) If $b=0$, we are left with a real number

(c) Therefore $a+ib$ represents all numbers real and imaginary.

(d) The quantity $a+ib$ is also known as a complex number. A complex has two parts to it - a real part and an imaginary part.

(e) The quantity $a+ib$ and $a-ib$ are called conjugates. The sum and product of conjugates is always real.

(f) It can be quickly seen that $\sqrt{a^2+b^2} = a\pm ib$.

(g) The modulus of a complex number $a+ib = \sqrt{a^2+b^2}$.

(h) The modulus of the product of two complex numbers is equal to the product of their moduli.

$$(a+ib)\times(c+id) = (ac-bd)+i(ad+bc).$$

The modulus of the product $=$

$$\sqrt{(ac-bd)^2 + (ad+bc)^2}$$

$$=\sqrt{a^2c^2 + b^2d^2 + a^2d^2 + b^2c^2}$$

$$=\sqrt{(a^2+b^2)(c^2+d^2)}$$

$$=\sqrt{a^2+b^2} \times \sqrt{c^2+d^2}$$

$=$ Product of the moduli of each of the complex numbers.

8. Let us turn our attention to the task of finding the square root of an imaginary expression $a+ib$

Let us assume that $\sqrt{a+ib} = x+iy$, where x and y are real quantities.

Squaring both sides of the equation, we get,

$$a+ib = x^2 - y^2 + 2ixy$$

Equating the real and imaginary parts, we get,

$$a = x^2 - y^2 \ldots(1)$$

$$b = 2xy \ldots(2)$$

$$(x^2+y^2)^2 = (x^2-y^2)^2 + (2xy)^2 = a^+b^2 \ldots(3)$$

$$x^2 + y^2 = \sqrt{a^2+b^2}$$

From equations (1) and (3), we can deduce,

$$x^2 = \frac{\sqrt{a^2+b^2} + a}{2}$$

$$y^2 = \frac{\sqrt{a^2+b^2} - a}{2}$$

Therefore, the values of x and y can be determined from these equations.

$$x = \pm\sqrt{\frac{\sqrt{a^2+b^2}+a}{2}}$$

$$y = \pm\sqrt{\frac{\sqrt{a^2+b^2}-a}{2}}$$

9. Let us now determine the cube roots of unity. In other words, we will solve for $x^3 - 1 = 0$ and determine the three roots of this cubic equation.

The first step is to factorize the cubic expression.

$$x^3 - 1 = 0$$

$$= (x-1)(x^2 + x + 1) = 0$$

Therefore $x - 1 = 0$ or $x = 1$ is one solution.

The other two solutions come from the roots of the quadratic expression $x^2 + x + 1 = 0$.

The two roots are $x = \dfrac{-1 \pm \sqrt{3}}{2}$

Let us denote these roots by the symbols α and β.

From theory of quadratic equations, we know that the constant term in the quadratic is simply the product of the roots.

$$\therefore \alpha \times \beta = 1$$

$$\alpha^3 \beta = \alpha^2$$

Since $\alpha^3 = 1$, we can conclude that $\beta = \alpha^2$.

Similarly, we can deduce that $\alpha = \beta^2$

Since each of the imaginary roots of this expression is square of the other, we denote the cube roots of unity as $1, \omega, \omega^2$. This leads us to the following observations

(a) $1+\omega+\omega^2 = 0$. This means sum of the cube roots of unity is zero.

(b) Product of the imaginary roots is unity. $\omega \times \omega^2 = 1$

(c) Every integral power of ω^3 is unity. Or $\omega^{3n} = 1$

10.1 Solved problems

1. $\dfrac{1}{1+\sqrt{2}-\sqrt{3}}$

$$\frac{1}{1+\sqrt{2}-\sqrt{3}} = \frac{1}{\left(1+\sqrt{2}-\sqrt{3}\right)}\frac{\left(1+\sqrt{2}+\sqrt{3}\right)}{\left(1+\sqrt{2}+\sqrt{3}\right)}$$

$$= \frac{1\sqrt{2}+\sqrt{3}}{\left(1+\sqrt{2}\right)^2 - \left(\sqrt{3}\right)^2}$$

$$= \frac{1+\sqrt{2}+\sqrt{3}}{1+2\sqrt{2}+2-3} = \frac{1+\sqrt{2}+\sqrt{3}}{2\sqrt{2}} \times \frac{\sqrt{2}}{\sqrt{2}}$$

$$= \frac{2+\sqrt{2}+\sqrt{6}}{4}$$

2. $\dfrac{\sqrt{2}}{\sqrt{2}+\sqrt{3}-\sqrt{5}}$

$$\frac{\sqrt{2}}{\sqrt{2}+\sqrt{3}-\sqrt{5}} = \frac{\sqrt{2}\left(\sqrt{2}+\sqrt{3}+\sqrt{5}\right)}{\left(\sqrt{2}+\sqrt{3}-\sqrt{5}\right)\left(\sqrt{2}+\sqrt{3}+\sqrt{5}\right)}$$

$$= \frac{2+\sqrt{6}+\sqrt{10}}{2+3+2\sqrt{6}-5} = \frac{2+\sqrt{6}\times\sqrt{10}}{2\sqrt{6}} \times \frac{\sqrt{6}}{\sqrt{6}}$$

$$= \frac{2\sqrt{6}+6+2\sqrt{15}}{12} = \frac{3+\sqrt{6}+\sqrt{15}}{6}$$

2.

$$\frac{1}{\sqrt{a}+\sqrt{b}+\sqrt{a+b}}$$

$$\frac{1}{\sqrt{a}+\sqrt{b}+\sqrt{a+b}}$$

$$=\frac{\sqrt{a}+\sqrt{b}-\sqrt{a+b}}{\left(\sqrt{a}+\sqrt{b}+\sqrt{a+b}\right)\left(\sqrt{a}+\sqrt{b}-\sqrt{a+b}\right)}$$

$$=\frac{\sqrt{a}+\sqrt{b}-\sqrt{a+b}}{a+b+2\sqrt{ab}-(a+b)}=\frac{\sqrt{a}+\sqrt{b}-\sqrt{a+b}}{2\sqrt{ab}}\times\frac{\sqrt{ab}}{\sqrt{ab}}$$

$$=\frac{a\sqrt{b}+b\sqrt{a}-\sqrt{ab(a+b)}}{2ab}$$

3.

$$\frac{2\sqrt{a+1}}{\sqrt{a-1}-\sqrt{2a}+\sqrt{a+1}}$$

$$\frac{2\sqrt{a+1}}{\sqrt{a+1}+\sqrt{a-1}-\sqrt{2a}}\times\frac{\left(\sqrt{a+1}+\sqrt{a-1}+\sqrt{2a}\right)}{\left(\sqrt{a+1}+\sqrt{a-1}+\sqrt{2a}\right)}$$

$$=\frac{2\left(a+1+\sqrt{a^2-1}+\sqrt{2a(a+1)}\right)}{a+1+a-1+2\sqrt{a^2-1}-2a}$$

$$=\frac{\cancel{2}\left(a+1+\sqrt{a^2-1}+\sqrt{2a^2+2a}\right)}{\cancel{2}\sqrt{a^2-1}}\times\frac{\sqrt{a^2-1}}{\sqrt{a^2-1}}$$

$$=\frac{(a+1)\sqrt{a^2-1}+a^2-1+\sqrt{2a(a+1)(a^2-1)}}{a^2-1}$$

$$=\frac{(a+1)\sqrt{a^2-1}+(a+1)(a-1)+\sqrt{2a(a+1)(a+1)(a-1)}}{(a+1)(a-1)}$$

$$= \frac{\sqrt{a^2-1}+a-1+\sqrt{2a(a-1)}}{a-1}$$

4. $\dfrac{\sqrt{10}+\sqrt{5}-\sqrt{3}}{\sqrt{3}+\sqrt{10}-\sqrt{5}}$

$$\frac{\sqrt{10}+\sqrt{5}-\sqrt{3}}{\sqrt{3}+\sqrt{10}-\sqrt{5}} = \frac{\sqrt{10}+\sqrt{5}-\sqrt{3}}{\sqrt{10}-\left(\sqrt{5}-\sqrt{3}\right)} \times \frac{\sqrt{10}+\sqrt{5}-\sqrt{3}}{\sqrt{10}+\left(\sqrt{5}-\sqrt{3}\right)}$$

$$= \frac{10+5+2\sqrt{50}+3-2\sqrt{3}\left(\sqrt{10}+\sqrt{5}\right)}{10-\left(5+3-2\sqrt{15}\right)}$$

$$= \frac{18+10\sqrt{2}-2\sqrt{30}-2\sqrt{15}}{2+2\sqrt{15}} = \frac{9+5\sqrt{2}-\sqrt{30}-\sqrt{15}}{1+\sqrt{15}}$$

$$= \frac{\left(9+5\sqrt{2}-\sqrt{15}-\sqrt{30}\right)\left(\sqrt{15}-1\right)}{\left(\sqrt{15}+1\right)\left(\sqrt{15}-1\right)}$$

$$= \frac{9\sqrt{15}+5\sqrt{30}-15-15\sqrt{2}-9-5\sqrt{2}+\sqrt{5}+\sqrt{30}}{14}$$

$$= \frac{-24-20\sqrt{2}+10\sqrt{15}+6\sqrt{30}}{14}$$

$$= \frac{-12-10\sqrt{2}+5\sqrt{15}+3\sqrt{30}}{7}$$

5. $\sqrt[3]{5}-\sqrt[4]{3}$

$5^{\frac{1}{3}}-3^{\frac{1}{4}}$; LCM of 3 and 4 = 12

Factor $= 5^{\frac{11}{3}}+5^{\frac{10}{3}}\cdot3^{\frac{1}{4}}+5^{\frac{9}{3}}3^{\frac{2}{4}}+5^{\frac{8}{3}}3^{\frac{3}{4}}+5^{\frac{7}{3}}3+5^23^{\frac{5}{4}}$

$\qquad +5^{\frac{5}{3}}3^{\frac{6}{4}}+5^{\frac{4}{3}}3^{\frac{7}{4}}+5\cdot3^2+5^{\frac{2}{3}}3^{\frac{9}{4}}$

$$+5^{1/3}3^{10/4}+3^{11/4}$$

6. $\dfrac{\sqrt[3]{3}-1}{\sqrt[3]{3}+1}$

To rationalize the denominator, the

Factor is $\dfrac{3^{2/3}-3^{1/3}+1}{2}$

$$\left(\dfrac{3^{1/3}-1}{3^{1/3}+1}\right)\left(\dfrac{3^{2/3}-3^{1/3}+1}{3^{2/3}-3^{1/3}+1}\right)=\dfrac{3-3^{2/3}+3^{1/3}-3^{2/3}+3^{1/3}-1}{3^{3/3}+1^3}$$

$$=\dfrac{2\cdot3^{1/3}-2\cdot3^{2/3}+2}{4}=\dfrac{3^{1/3}-3^{2/3}+1}{2}$$

7. $\left(135\sqrt{3}-87\sqrt{6}\right)^{\frac{1}{3}}$

$$\left(135\sqrt{3}-87\sqrt{6}\right)=\left(45\cdot3\sqrt{3}-29\cdot3\sqrt{3}\cdot\sqrt{2}\right)^{1/3}$$

$$=\sqrt{3}\left(45-29\sqrt{2}\right)^{1/3}$$

Let $\left(45-29\sqrt{2}\right)^{1/3}=x-\sqrt{y}\Rightarrow\left(45+29\sqrt{2}\right)^{1/3}=x+\sqrt{y}$

$$\Rightarrow x^2-y=\left(45^2-29\times27\right)^{1/3}=(2025-1682)^{1/3}=343^{1/3}$$

$$\Rightarrow x^2-y=7$$

Also, $x^3+3xy=45;\sqrt{y}\left(3x^2+y\right)=29\sqrt{2}$

If $y=2=x=\sqrt{7+2}=3\Rightarrow3x^2+y=27+2=29$

Root $=\sqrt{3}\left(3-\sqrt{2}\right)=3\sqrt{3}-\sqrt{6}$

8. $\sqrt{\dfrac{6+2\sqrt{3}}{33-19\sqrt{3}}}$

$$\sqrt{\frac{6+2\sqrt{3}}{33-19\sqrt{3}}} = \sqrt{\frac{\left(1+2\sqrt{3}\right)\left(33+19\sqrt{3}\right)}{\left(33-19\sqrt{3}\right)\left(33+19\sqrt{3}\right)}}$$

$$= \sqrt{\frac{198+66\sqrt{3}+114\sqrt{3}+144}{6}}$$

$$= \sqrt{\frac{312+180\sqrt{3}}{6}} = \sqrt{52+30\sqrt{3}} = \sqrt{x}+\sqrt{y}$$

$$\Rightarrow x+y = 52; 2\sqrt{xy} = 30\sqrt{3} \Rightarrow 4xy = 2700$$

$$\Rightarrow x-y = \sqrt{(x+y)^2 - 4xy} = \sqrt{52^2 - 2700} = \sqrt{4} = 2$$

$$x = 27; y = 25 \Rightarrow \text{Ans} = \sqrt{27}+\sqrt{25}$$

$$= 3\sqrt{3}+5$$

9. $\left(26+15\sqrt{3}\right)^{\frac{2}{3}} - \left(26+15\sqrt{3}\right)^{-\frac{2}{3}}$

Let $\left(26+15\sqrt{3}\right)^{\frac{1}{3}} = x+\sqrt{y}$

$$\Rightarrow \left(26-15\sqrt{3}\right)^{\frac{1}{3}} = x-\sqrt{y}$$

$$\Rightarrow x^2 - y = \left(26^2 - 15^2 \cdot 3\right)^{\frac{1}{3}} = 1 \Rightarrow x^2 = y+1$$

Cubing, $26+15\sqrt{3} = \left(x+\sqrt{y}\right)^3$

$$\Rightarrow x^3 + 3xy = 26$$

$$\sqrt{y}\left(3x^2 + y\right) = 15\sqrt{3}$$

If $y = 3; x = \sqrt{y+1} = 2 \Rightarrow 3x^3 + y = 15; x^3 + 3xy = 26$

Hence $\left(26 + 15\sqrt{3}\right)^{1/3} = 2 + \sqrt{3}$

$$\Rightarrow \left(26 + 15\sqrt{3}\right)^{2/3} = 4 + 3 + 4\sqrt{3} = 7 + 4\sqrt{3}$$

Ans. $= 7 + 4\sqrt{3} - \dfrac{1}{7 + 4\sqrt{3}} = 7 + 4\sqrt{3} - \left(\dfrac{7 - 4\sqrt{3}}{49 - 48}\right) = 8\sqrt{3}$

10.   **Given $\sqrt{5} = 2.23607$, find the value of**

$$\dfrac{10\sqrt{2}}{\sqrt{18} - \sqrt{3 + \sqrt{5}}} - \dfrac{\sqrt{10} + \sqrt{18}}{\sqrt{8} + \sqrt{3 - \sqrt{5}}}$$

Let $\sqrt{3 + \sqrt{5}} = \sqrt{x} + \sqrt{y}$

$$\Rightarrow 3 + \sqrt{5} = x + y + 2\sqrt{xy}$$

$$x + y = 3; 4xy = 5$$

$$\Rightarrow (x - y) = \sqrt{(x+y)^2 - 4xy} = \sqrt{9 - 5} = 2$$

$$x = \dfrac{5}{2}; y = \dfrac{1}{2}$$

$$\sqrt{3 + \sqrt{5}} = \dfrac{\sqrt{5} + 1}{\sqrt{2}}$$

$$\sqrt{3 - \sqrt{5}} = \dfrac{\sqrt{5} - 1}{\sqrt{2}}$$

Therefore,

$$= \dfrac{10\sqrt{2}}{\sqrt{18} - \dfrac{\sqrt{5}+1}{\sqrt{2}}} - \dfrac{\sqrt{10} + \sqrt{18}}{\sqrt{8} + \dfrac{\sqrt{5}-1}{\sqrt{2}}} = \dfrac{10 \times 2}{\sqrt{36} - \left(\sqrt{5}+1\right)} - \dfrac{\sqrt{20} + \sqrt{36}}{\sqrt{16} + \left(\sqrt{5}-1\right)}$$

$$= \frac{20}{6-\sqrt{5}-1} - \frac{2\sqrt{5}+6}{4+\sqrt{5}-1} = \frac{20}{5-\sqrt{5}} - \frac{2\sqrt{5}+6}{3+\sqrt{5}}$$

$$= \frac{20\left(5+\sqrt{5}\right)}{\left(5+\sqrt{5}\right)\left(5-\sqrt{5}\right)} - \frac{\left(2\sqrt{5}+6\right)\left(3-\sqrt{5}\right)}{\left(3+\sqrt{5}\right)\left(3-\sqrt{5}\right)}$$

$$= 5+\sqrt{5} - \frac{6\sqrt{5}+18-10-6\sqrt{5}}{4}$$

$$= 5+\sqrt{5}-2 = 3+\sqrt{5} = 5\cdot 23607$$

11. **Multiply** $2\sqrt{-3}+3\sqrt{-2}$ **by** $4\sqrt{-3}-5\sqrt{-2}$.

$$\left(2\sqrt{3}i+3\sqrt{2}i\right)\times\left(4\sqrt{3}i-5\sqrt{2}i\right)$$

$$= 8\cdot 3i^2 + 12\sqrt{b}i^2 - 10\sqrt{6}i^2 - 15\cdot 2i^2$$

$$= -24-2\sqrt{6}+10\sqrt{6}+30$$

$$= 6-2\sqrt{6}$$

12. **Multiply** $e^{\sqrt{-1}}+e^{-\sqrt{-1}}$ **by** $e^{\sqrt{-1}}-e^{-\sqrt{-1}}$

$$\left(e^i+e^{-i}\right)\times\left(e^i-e^{-i}\right)$$

$$= e^{i+i}+e^{-i+i}-e^{i+-i}-e^{-i-i}$$

$$= e^{2i}-e^{2i}$$

13. **Multiply** $x-\dfrac{1+\sqrt{-3}}{2}$ **by** $x-\dfrac{1-\sqrt{-3}}{2}$

$$\left(x-\frac{1+\sqrt{3}i}{2}\right)\times\left(x-\frac{1-\sqrt{3}i}{2}\right)$$

$$= x^2 - \left[\frac{1+\sqrt{3}i}{2}+\frac{1-\sqrt{3}i}{2}\right]x + \left(\frac{1+\sqrt{3}i}{2}\right)\left(\frac{1-\sqrt{3}i}{2}\right)$$

$$= x^2 - x + \frac{1-(-3)}{4} = x^2 - x + 1$$

Express with rational denominator:

14. $\dfrac{1}{3-\sqrt{-2}}$

$$\frac{1}{3-\sqrt{2}i} = \frac{1 \times \left(3+\sqrt{2}i\right)}{\left(3-\sqrt{2}i\right)\left(3+\sqrt{2}i\right)} = \frac{3+\sqrt{2}i}{9+2} = \frac{3+\sqrt{2}i}{11}$$

15.   **Find the value of** $\left(-\sqrt{-1}\right)^{4n+3}$ **when n is a positive integer.**

$$\left(-i\right)^{4n+3} = (-1)^{4n+3}(i)^{4n+3}$$

$$= (-1)^{4n} \cdot (-1)^3 \cdot (i)^{4n} \cdot i^3$$

$$= (-1^2)^{2n} \cdot (-1)^3 \cdot (i^4)^n \cdot i^3 \quad \begin{vmatrix} -1^2 = 1 \\ i^n = (i^2)^2 = (-1)^2 = 1 \end{vmatrix}$$

$$= (-1)^3 \cdot i^3 = -1 \cdot -1 \cdot i$$

$$= i$$

16. $\sqrt{-47+8\sqrt{-3}}$

$$\sqrt{-47+8\sqrt{3}i} = x + iy$$

$$-47+8\sqrt{3}i = \left(x^2 - y^2\right)72ixy$$

$$x^2 - y^2 = -47; \ 2xy = 8\sqrt{3}$$

$$\left(x^2+y^2\right)^2 = \left(x^2-y^2\right)^2 + 4x^2y^2 = (-47)^2 + \left(8\sqrt{3}\right)^2$$

$$= 2401 = 49^2$$

$$\Rightarrow x^2 + y^2 = 49$$

$$\Rightarrow x^2 = 1; y^2 = 48 = 16 \times 3$$

$$\Rightarrow x = \pm 1; \ y = \pm 4\sqrt{3}$$

xy is positive $\Rightarrow$ x, y have the same sign

$$\text{Answer} = \pm\left(1 + 4\sqrt{3}i\right)$$

17. $\sqrt{-8\sqrt{-1}}$

$$\sqrt{-8i} = x + iy$$

$$\Rightarrow x^2 - y^2 = 0; 2xy = -8$$

$$x = \pm y \Rightarrow xy = -4$$

$$\Rightarrow x = 2, y = -2 \text{ or } x = -2, y = 2$$

$$\sqrt{-8i} = \pm\left(2 - 2i\right)$$

18. $\sqrt{a^2 - 1 + 2a\sqrt{-1}}$,

$$\sqrt{a^2 - 1 + 2ai} = \sqrt{(a+i)^2} = \pm(a+i)$$

Alternate method:

$$\sqrt{a^2 - 1 + 2ai} = x + iy$$

$$\Rightarrow x^2 - y^2 = a^2 - 1; \ 2xy = 2a$$

$$x^2 + y^2 = \sqrt{(x^2 - y^2) + 4x^2 y^2} = \sqrt{(a^2 - 1)^2 + 4a^2}$$

$$= \sqrt{a^4 - 2a^2 + 1 + 4a^2}$$

$$= \sqrt{a^4 + 2a^2 + 1} = a^2 + 1$$

$$x = a^2; y = 1 \Rightarrow x = \pm a; y = \pm 1$$

xy is positive $\Rightarrow$ x, y have the same sign

$$\text{Answer} = \pm\left(a + i\right)$$

19. $4ab - 2(a^2 - b^2)\sqrt{-1}$

$\sqrt{4ab - 2(a^2 - b^2)i} = x - iy$

$\Rightarrow x^2 - y^2 = 4ab$

$2xy = 2\left(a^2 - b^2\right)$

$x^2 + y^2 = \sqrt{(x^2 - y^2)^2 + 4x^2 y^2} = \sqrt{16a^2 b^2 + 4(a^2 - b^2)^2}$

$= 2\sqrt{4a^2 b^2 + a^4 + b^4 - 2a^2 b^2} = 2\sqrt{a^4 + b^4 + 2a^2 b^2}$

$= 2\left(a^2 + b^2\right)$

$\Rightarrow x^2 = a^2 + b^2 + 2ab \Rightarrow x = \pm(a + b)$

$y^2 = a^2 + b^2 - 4ab \Rightarrow y = \pm(a - b)$

xy is positive so x, y have the same sign

Answer $= \pm\left(a + b - (a - b)i\right)$

20. $\dfrac{3 + 5i}{2 - 3i}$

$\dfrac{3 + 5i}{2 - 3i} = \dfrac{(3 + 5i)(2 + 3i)}{(2 - 3i)(2 + 3i)} = \dfrac{6 - 15 + 10i + 9i}{4 + 9}$

$= \dfrac{-9 + 19i}{13} = \dfrac{-9}{13} + \dfrac{19}{13}i = \dfrac{-9}{13} + \dfrac{19}{13}i$

21. $\dfrac{\sqrt{3} - i\sqrt{2}}{2\sqrt{3} - i\sqrt{2}}$

$\dfrac{\sqrt{3} - \sqrt{2}i}{2\sqrt{3} - \sqrt{2}i} = \dfrac{\left(\sqrt{3} - \sqrt{2}i\right)\left(2\sqrt{3} + \sqrt{2}i\right)}{\left(2\sqrt{3} - \sqrt{2}i\right)\left(2\sqrt{3} + \sqrt{2}i\right)}$

$= \dfrac{2 \times 3 - 2\sqrt{6}i + \sqrt{6}i - 2i^2}{4 \times 3 + 2} = \dfrac{8 - \sqrt{6}i}{14} = \dfrac{4}{7} - \dfrac{\sqrt{6}i}{14}$

11 Quadratic and Miscellaneous Equations

1. The general representation of a quadratic equation is $ax^2 + bx + c = 0$.

2. The solution of the equations is

$$x = \frac{-b \pm \sqrt{b^2 - 4ac}}{2a}.$$

The solutions of the quadratic equation are also known as the root of the quadratic equation. We use the term roots and solutions interchangeably.

3. We can make a set of observations about the roots of the quadratic equation.

4. A quadratic equation cannot have more than two roots.

Proof: Let us assume that the quadratic equation had three roots α, β, γ. If this is the case, then we have the following equations:

$$a\alpha^2 + b\alpha + c = 0 \ldots (1)$$

$$a\beta^2 + b\beta + c = 0 \ldots (2)$$

$$a\gamma^2 + b\gamma + c = 0 \ldots (3)$$

Subtracting (2) from (1), we get:

$$a(\alpha^2 - \beta^2) + b(\alpha - \beta) = 0 \ldots (4)$$

This means, either

$\alpha = \beta$ [This is not possible, since we have assumed that the equation has 3 roots] or

$$\alpha + \beta = -\frac{b}{a} \ldots (5)$$

By manipulating (1) and (3), we get

$$a(\alpha^2 - \gamma^2) + b(\alpha - \gamma) = 0 \ldots (4)$$

This means, either

$\alpha = \gamma$ [This is not possible, since we have assumed that the equation has 3 roots] or

$$\alpha + \gamma = -\frac{b}{a} \ \ldots (6)$$

Equations (5) = (6).

$\therefore$, it follows that $\beta = \gamma$.

In other words, we can have only two solutions to a quadratic equation.

5. **We have seen that the roots of the quadratic equation can be determined from the expression**

$$\frac{-b \pm \sqrt{b^2 - 4ac}}{2a}$$

(i) $b^2 - 4ac = 0$

In this case, the roots are real and equal to one another. The quadratic is simply a square of an algebraic expression; and is of the form $(x \pm \alpha)^2 = 0$.

(j) $b^2 - 4ac > 0$

In this case, the roots are real and unequal. The quadratic expression is of the form $(x \pm \alpha)(x \pm \beta) = 0$.

6. $b^2 - 4ac < 0$

In this case, the roots are imaginary and unequal. And the roots are of the form $m \pm in$ where $i = \sqrt{-1}$.

7. $b^2 - 4ac$ is a perfect square

In this case, the roots are rational and unequal.

8. Let us look at the roots of the quadratic equation one more time.

$$\alpha = \frac{-b+\sqrt{b^2-4ac}}{2a}\ldots(1)$$

$$\beta = \frac{-b-\sqrt{b^2-4ac}}{2a}\ldots(2)$$

$$\therefore \alpha+\beta = \frac{-b+\sqrt{b^2-4ac}}{2a}+\frac{-b-\sqrt{b^2-4ac}}{2a}$$

$$= \frac{-b+\sqrt{b^2-4ac}-b-\sqrt{b^2-4ac}}{2a}$$

$$= -\frac{b}{a}\ldots(3)$$

And $\alpha \times \beta$

$$= \frac{(-b+\sqrt{b^2-4ac})(-b-\sqrt{b^2-4ac})}{2a}$$

$$= \frac{(-b)^2-(b^2-4ac)}{4a^2}$$

$$= \frac{c}{a}$$

Therefore, the quadratic equation can be represented as follows :

$$ax^2+bx+c=0\ldots(1)$$

$$x^2+\frac{b}{a}x+\frac{c}{a}=0\ldots(2)$$

$$x^2-x\times(sumofroots)+(productoftheroots)=0\ldots(4)$$

9. The condition for the roots of a quadratic equation to be equal in magnitude and opposite in **sign** is $\alpha + \beta = 0$ or $-\dfrac{b}{a} = 0$ or $b = 0$

10. The condition for the roots to be reciprocal of one another is $\alpha \times \beta = 1$ or $\dfrac{c}{a} = 1$ or $c = a$.

11.1 Solved problems

Solve the following equations:

1. $x^{-2} - 2x^{-1} = 8$

$$x^{-2} - 2x^{-1} = 8 \Rightarrow 1 - 2x = 8x^2 \ (x^2)$$

Or $8x^2 + 2x - 1 = 0$

$$(4x - 1)(2x + 1) = 0$$

$$x = \frac{1}{4} \text{ or } \frac{-1}{2}$$

2. $2\sqrt{x} + 2x^{-\frac{1}{2}} = 5$

Put $\sqrt{x} = y \Rightarrow x = y^2$

$$2y + \frac{2}{y} = 5 \Rightarrow 2y^2 + 2 = 5y$$

$$2y^2 - 5y + 2 = 0 \Rightarrow (2y - 1)(y - 2\_ = 0$$

$$y = 2 \text{ or } \frac{1}{2}$$

$$x = y^2 = 4 \text{ or } \frac{1}{4}$$

3. $6x^{\frac{3}{4}} = 7x^{\frac{1}{4}} - 2x^{-\frac{1}{4}}$

Multiply by $x^{1/4}$

$$\Rightarrow 6x = 7x^{1/2} - 2$$

Put $x^{1/2} = y \Rightarrow 6y^2 - 7y + 2 = 0$

$$\Rightarrow (3y - 2)(2y - 1) = 0$$

$$y = \frac{2}{3} \text{ or } \frac{1}{2} \Rightarrow x = y^2 = \frac{4}{9} \text{ or } \frac{1}{4}$$

4. $x^{\frac{2}{n}} + 6 = 5x^{\frac{1}{n}}$

Put $x^{1/n} = y \Rightarrow y^2 + 6 = 5y \Rightarrow y^2 - 5y + 6 = 0$

$$\Rightarrow (y - 2)(y - 3) = 0$$

$$y = 2 \text{ or } 3 \Rightarrow x = y^n = 2^n \text{ or } 3^n$$

5. $3x^{\frac{1}{2n}} - x^{\frac{1}{n}} - 2 = 0$

Put $x^{1/2n} = y \Rightarrow x^{1/n} = y^2 \Rightarrow x = y^{2n}$

$$\Rightarrow 3y - y^2 - 2 = 0 \text{ or } y^2 - 3y + 2 = 0$$

$$(y - 1)(y - 2) = 0 \text{ or } y = 1 \text{ or } 2$$

$$x = y^{2x} = 1 \text{ or } 2^{2n}$$

6. $5\sqrt{\dfrac{3}{x}} + 7\sqrt{\dfrac{x}{3}} = 22\dfrac{2}{3}.$

Put $\sqrt{\dfrac{x}{3}} = y \Rightarrow x = 3y^2$

$$\Rightarrow 7y + \frac{5}{y} = \frac{68}{3} \Rightarrow 21y^2 + 15 = 68y$$

Or $21y^2 - 68y + 15 = 0$

$$y = 3 \text{ or } \frac{5}{21}$$

$$x = 3 y^2 = 27 \text{ or } \frac{25}{147}$$

7. $3^{2x} + 9 = 10.3^x$

Put $3^x = y \Rightarrow x \log 3 = \log y$ or $x = \dfrac{\log y}{\log 3}$

$$\Rightarrow y^2 + 9 = 10 y \Rightarrow y^2 - 10 y + 9 = 0$$

$$(y - 1)(y - 9) = 0 \Rightarrow y = 1 \text{ or } 9$$

$$x = \frac{\log 1}{\log 3} \text{ or } \frac{\log 9}{\log 3} = 0 \text{ or } 2 \left| \begin{array}{l} 9 = 3^2 \\ \log 9 = 2 \log 3 \\ 1 = 3 0 \end{array} \right.$$

8. $2^{2x+3} + 1 = 32.2^x$

Put $2^x = y \Rightarrow 256 \cdot y^2 + 1 - 32 y = 0$

$$\Rightarrow (16 y - 1)^2 = 0 \Rightarrow y = \frac{1}{16} = 2^{-4}$$

$$x = -4$$

9. $\dfrac{3x - 2}{2} + \sqrt{2x^2 - 5x + 3} = \dfrac{(x + 1)^2}{3}$

$$\Rightarrow 9x - 6 + 6\sqrt{2x^2 - 5x + 3} = 2x^2 + 4x + 2$$

$$\Rightarrow 2x^2 - 5x + 8 + 6\sqrt{2x^2 - 5x + 3} = 0$$

Put $y = \sqrt{2x^2 - 5x + 3}$

$$\Rightarrow y^2 + 6 y + 5 = 0 \Rightarrow (y + 1)(y + 5) = 0$$

$$y = -1 \text{ or } -5$$

$$2x^2 - 5x + 3 = y^2 = 1 \text{ or } 25$$

$$\Rightarrow 2x^2 - 5x + 2 = 0 \text{ Or } 2x^2 - 5x - 22 = 0$$

$$(2x-1)(x-2) = 0 \text{ or } x = \frac{5 \pm \sqrt{201}}{4}$$

$$x = \frac{1}{2}, 2, \frac{5 \pm \sqrt{201}}{4}$$

10. $x^4 + \dfrac{8}{9}x^2 + 1 = 3x^3 + 3x.$

$$\Rightarrow 9x^4 + 8x^2 + 9 = 27x^3 + 27x$$

$$\div \text{ by } x^2; 9x^2 + 8 + \frac{9}{x^2} - 27x - \frac{27}{x} = 0$$

$$\Rightarrow 9x^2 + \frac{9}{x^2} + 18 - 27\left(x + \frac{1}{x}\right) - 10 = 0$$

$$9\left(x + \frac{1}{x}\right)^2 - 27\left(x + \frac{1}{x}\right) - 10 = 0$$

Put $y = \dfrac{1}{x} + x \Rightarrow 9y^2 - 27y - 10 = 0$

$$x = \frac{27 \pm \sqrt{1089}}{18} = \frac{27 \pm 33}{18} = \frac{10}{3} \text{ or } \frac{-1}{3}$$

$$\Rightarrow x + \frac{1}{x} = \frac{10}{3} \text{ or } \frac{-1}{3} \Rightarrow x^2 + 1 = \frac{10}{3}x \text{ or } \frac{-1}{3}x$$

Or $3x^2 - 10x + 3 = 0$ or $3x^2 + x + 3 = 0$

$$x = \frac{10 \pm \sqrt{64}}{6} \text{ Or } \frac{-1 \pm \sqrt{1-36}}{6}$$

$$= \frac{10 \pm 8}{6} \text{ Or } \frac{-1 \pm \sqrt{35}i}{6}$$

$$= 3, \frac{1}{3}, \frac{-1 \pm \sqrt{35}i}{6}$$

11. $x^4 - 2x^3 + x = 380$

$$x^4 - 2x^3 + x^2 - x^2 + x - 380 = 0$$

$$(x^2 - x)^2 - (20 - 19)(x^2 - x) - 20 \times 19 = 0$$

$$(x^2 - x - 20)(x^2 - x + 19) = 0$$

$$(x - 5)(x + 4)(x^2 - x + 19) = 0$$

$$x = 5, -4, \frac{1 \pm \sqrt{1 - 76}}{2} \quad |75 = 25 \times 3$$

$$= 5, -4, \frac{1 \pm 5\sqrt{3}i}{2}$$

Solve the following equations:

12. $3x - 2y = 7, xy = 20$.

$$\Rightarrow > y = \frac{20}{x}$$

$$\Rightarrow 3x - \frac{2 \times 20}{x} = 7 \Rightarrow 3x^2 - 40 = 7x$$

$$3x^2 - 7x - 40 = 0; \quad x = \frac{7 \pm 23}{6} = 5 \,\text{or}\, \frac{-8}{3}$$

$$y = \frac{20}{x} = 4 \,\text{Or}\, \frac{-15}{2}$$

$$(x, y) = (5, 4) \,\text{or}\, \left(-\frac{8}{3}, -\frac{15}{2}\right)$$

13. $5x - y = 3, y^2 - 6x^2 = 25$

$$y = 5x - 3 \Rightarrow (5x - 3)^2 \, 6x^2 = 25$$

$$\Rightarrow 25x^2 - 6x^2 - 30x + 9 = 25$$

$$\Rightarrow 19x^2 - 30x - 16 = 0$$

$$x = \frac{30 \pm \sqrt{2116}}{38} = \frac{30 \pm 46}{38} = 2 \text{ or } \frac{-8}{19}$$

$$y = 5x - 3 = 7 \text{ Or } \frac{-97}{19}$$

14. $4x - 3y = 1, 12xy + 13y^2 = 25$

$$\Rightarrow x = \frac{1 + 3y}{4} \Rightarrow 12y\left(\frac{1 + 3y}{4}\right) + 13y^2 = 25$$

$$\Rightarrow 3y + 9y^2 + 13y^2 = 25 \Rightarrow 22y^2 + 3y - 25 = 0$$

$$y = \frac{-3 \pm \sqrt{2209}}{44} = \frac{-3 \pm 47}{44} = 1 \text{ or } \frac{-25}{22}$$

$$x = \frac{1 + 3y}{4} = 1 \text{ Or } \frac{1 - \dfrac{75}{22}}{4} = \frac{-53}{88}$$

15. $x^4 + x^2 y^2 + y^4 = 931, x^2 - xy + y^2 = 19$.

$$x^4 + x^2 y^2 + y^4 = \left(x^2 + y^2 - xy\right)\left(x^2 + y^2 + xy\right)$$

$$\Rightarrow x^2 + y^2 + xy = \frac{931}{19} = 49$$

$$\Rightarrow xy = \frac{49 - 19}{2} = 15$$

$$\Rightarrow \left(x + y\right)^2 = x^2 + y^2 + 2xy = 49 + 15 = 64$$

$$x + y = \pm 8$$

$$\left(x - y\right)^2 = x^2 + y^2 - 2xy = 19 - 15 = 4$$

$$x - y = \pm 2$$

$$x = \frac{\pm 8 \pm 2}{2} = \pm 5 \, or \pm 3$$

$$y = \pm 3 \, or \pm 5$$

16. $x + \sqrt{xy} + y = 65, x^2 + xy + y^2 = 2275.$

$$\Rightarrow x + y - \sqrt{xy} = \frac{x^2 + y^2 + xy}{x + y + \sqrt{xy}} = \frac{2275}{65} = 35$$

$$\Rightarrow x + y = 50; \sqrt{xy} = 15 \Rightarrow xy = 225$$

$$x = \frac{225}{y} \Rightarrow y + \frac{225}{y} = 50$$

$$\Rightarrow y^2 - 50y + 225 = 0 \Rightarrow (y-5)(y-45) = 0$$

$$y = 5 \, or \, 45, x = 45 \, or \, 5$$

17. $x + y = 7 + \sqrt{xy}, x^2 + y^2 = 133 - xy.$

$$\Rightarrow x + y = -\sqrt{xy} = 7; x^2 + y^2 + xy = 133$$

$$x + y + \sqrt{xy} = \frac{133}{7} = 19$$

$$x + y = \frac{26}{2} = 13 \Rightarrow \sqrt{xy} = 6 \Rightarrow xy = 36$$

$$\Rightarrow x = \frac{36}{y} \, Or \; y + \frac{36}{y} = 13$$

$$\Rightarrow y^2 - 13y + 36 = 0 \Rightarrow (y-4)(y-9) = 0$$

$$y = 4 \, or \, 9; \; x = 9 \; or \, 4$$

18. $5y^2 - 7x^2 = 17, 5xy - 6x^2 = 6.$

Let $y = mx$

$$\Rightarrow x^2\left(5m^2 \cdot 7\right) = 17; x^2\left(5m - 6\right) = 6$$

$$\Rightarrow \frac{5m^2 - 7}{5m - 6} = \frac{17}{6} \Rightarrow 30m^2 - 42 = 85m - 102$$

$$\Rightarrow 30m^2 - 85m + 60 = 0 \Rightarrow 6m^2 - 17m + 12 = 0$$

$$m = \frac{17 \pm 1}{12} = \frac{3}{2} \text{ or } \frac{4}{3}$$

$$x^2 = \frac{6}{5m - 6} = \frac{6}{\dfrac{5 \times 3}{2} - 6} \text{ or } \frac{6}{\dfrac{5 \times 4}{3} - 6}$$

$$= \frac{12}{3} \text{ Or } \frac{18}{2} = 4 \text{ or } 9$$

$$x = \pm 2 \text{ Or } \pm 3$$

$$y = \pm 3 \text{ Or } \pm 4$$

19. $3x^2 + 165 = 16xy, 7xy + 3y^2 = 132$.

$$y = mx$$

$$\Rightarrow n^2(16m - 3) = 165; x^2(7m + 3m^2) = 132$$

$$\Rightarrow \frac{3m^2 + 7m}{16m - 3} = \frac{132}{165} = \frac{4}{5}$$

$$\Rightarrow 15m^2 + 35m = 64m - 12$$

Or $15m^2 - 29m + 12 = 0$

$$m = \frac{29 \pm 11}{30} = \frac{40}{30} \text{ or } \frac{18}{30} = \frac{4}{3} \text{ or } \frac{3}{5}$$

$$x^2 = \frac{165}{16m - 3} = \frac{165}{16 \times \dfrac{4}{3} - 3} \text{ or } \frac{165}{16 \times \dfrac{3}{5} - 3}$$

$$= \frac{165 \times 3}{69 - 9} \text{ or } \frac{165 \times 5}{48 - 15} = \frac{165 \times 3}{55} \text{ or } \frac{165 \times 5}{33}$$

$$= 9 \text{ or } 25 \Rightarrow x = \pm 3 \text{ or } \pm 5$$

$$y = \pm 4 \text{ or } \pm 3$$

20. $x^4 + y^4 = 272, x - y = 2$

Let $x = m + n; \ y = m - n$

$\Rightarrow n = 1 \text{ or } x = m + 1; \ y = m - 1$

$x^4 + y^4 = (m + 1)^4 + (m - 1)^4 = 272$

$\Rightarrow (m^2 + 1)^2 + (2m^2)^2 = \dfrac{272}{2} = 136$

$\Rightarrow m^4 + 6m^2 - 135 = 0$

$(m^2 - 9)(m^2 + 15) = 0$

$m = \pm 3; \pm \sqrt{15}i$

$x = 4, -2, 1 + \sqrt{15}i, 1 - \sqrt{15}i$

$y = 2, -3, 1 + \sqrt{15}i, -1 - \sqrt{15}i$

21. $x^5 - y^5 = 992, x - y = 2$.

Let $x = m + n; \ y = m - n$

$\Rightarrow n = 1; n = m + 1; \ y = m - 1$

$x^5 - y^5 = (m + 1)^5 - (m - 1)^5$

$= m(m + 1)^4 + (m + 1)^4 - m(m - 1)^4 + (m - 1)^4$

$= m\left[m^4 + 2m + 1\right]^2 + \left((m^2 + 1)^2 + 4m^2\right) \times 2 - m\left[(m^2 + 1) - 2m\right]$

$$= m\left[\left(m^2+1\right)^1+4m\left(m^2+1\right)+4m^2-\left(m^2+1\right)^2-4m^2\right.$$

$$\left.+4m\left(m^2+1\right)\right]+2\left(m^4+6m^2+1\right)$$

$$= 8\underline{m}^2\left(m^2+1\right)+2\left(\underline{m}^4+6m^2+1\right)=992=496\times2$$

$$\Rightarrow 5m^4+10m^2=495 \Rightarrow m^4+2m^2=99$$

$$m^4+2m^2-99=0 \Rightarrow \left(m^2-9\right)\left(m^2+11\right)=0$$

$$m=\pm3;\pm\sqrt{11}i$$

$$x=4,-2,1+\sqrt{11}i,1-\sqrt{11}i$$

$$y=2,-4,-1+\sqrt{11}i,-1-\sqrt{11}i$$

22. $9x+y-8z=0, 4x-8y+7z=0, yz+zx+xy=47$

Cross multiplying; the first two equations

$$\begin{vmatrix} x & y & z \\ 9 & 1 & -8 \\ 4 & -8 & 7 \end{vmatrix}$$

$$\frac{x}{7-64}=\frac{y}{-32-63}=\frac{z}{-72-4}=k_1$$

$$\frac{x}{57}=\frac{y}{95}=\frac{z}{75}=k_2 \Rightarrow \frac{x}{3}=\frac{y}{5}=\frac{z}{4}=k_3$$

$$\Rightarrow x=3k; y=5k; z=4k$$

$$\Rightarrow (15+12+20)k^2=47 \Rightarrow k=\pm1$$

$$\Rightarrow x=\pm3; y=\pm5; z=\pm4$$

23. $3x+y-2z=0, 4x-y-3z=0, x^3+y^3+z^3=467.$

Cross multiplying the first two equations

$$\begin{vmatrix} x & y & z \\ 3 & 1 & -2 \\ 4 & -1 & 3 \end{vmatrix}$$

$$\Rightarrow \frac{x}{-3-2} = \frac{y}{-8+9} = \frac{z}{-3-4} = k$$

$$\Rightarrow x = 5k; y = k; z = -7k$$

$$\Rightarrow (-125 + 1 - 343)k^3 = 467$$

$$\Rightarrow k^3 = -1 \Rightarrow k = -1$$

$$x = 5, y = -1, z = 7$$

24. $x - y - z = 2; x^2 + y^2 - z^2 = 22, xy = 5.$

$$\Rightarrow x^2 + y^2 - 2xy - z^2 = 22 - 10 = 12$$

$$\Rightarrow (x-y)^2 - z^2 = 12 \Rightarrow (x - y - z)(x - y + z) = 12$$

$$\Rightarrow x - y + z = 6 \Rightarrow x - y = 4 \Rightarrow z = 2$$

$$\Rightarrow x - \frac{5}{x} = 4 \Rightarrow x^2 - 4x - 5 = 0$$

$$\Rightarrow (x-5)(x+1) = 0 \Rightarrow x = 5, -1; y = 1, -5$$

$$x = 5, y = 1, z = 2 \text{ Or } x = -1, y = -5, z = 2$$

25. $x + 2y - z = 11, x^2 - 4y^2 + z^2 = 37, xz = 24.$

$$x^2 - 4y^2 + z^2 = 37$$

$$x^2 + z^2 - 2xz - 4y^2 = 37 - 48 = -11$$

$$(x-z)^2 - 4y^2 = -11 \Rightarrow (x - z + 2y)(x - z - 2y) = -11$$

$$\Rightarrow x - z - 2y = -1$$

$$\Rightarrow x - z = 5; \ y = 3$$

$$\Rightarrow x - \frac{24}{x} = 5 \Rightarrow x^2 - 5x - 24 = 0$$

$$(x+3)(x-8) = 0 \Rightarrow x = 8 \text{ or} -3; \ z = 3 \text{ or} -8$$

$$(x, y, z) = (8, 3, -3) \text{ or} (-3, 3, -8)$$

26. $x^2 + y^2 - z^2 = 21, 3xz + 3yz - 2xy = 18, x + y - z = 5.$

$$\Rightarrow x + y = 5 + z \Rightarrow x^2 + 2xy + y^2 = z^2 + 10z + 25$$

$$\Rightarrow 21 + z^2 + 2xy = z^2 + 10z + 25$$

$$\Rightarrow 2xy = 10z + 4 \Rightarrow xy = 5z + 2$$

$$3z(x + y) - 2xy = 18$$

$$\Rightarrow 3z(z + 5) - 10z - 4 = 18$$

$$\Rightarrow 3z^2 + 5z - 22 = 0$$

$$z = \frac{-5 \pm \sqrt{25 + 264}}{6} = \frac{-5 \pm 17}{6} = 2 \text{ or} \frac{-11}{3}$$

(a) $z = 2 \Rightarrow x + y = 7; \ xy = 12$

$$\Rightarrow x + \frac{12}{x} = 7 \Rightarrow x^2 - 7x + 12 = 0 \Rightarrow (x-4)(x-3) = 0$$

$$\Rightarrow x = 4; y = 3 \text{ Or } x = 3, y = 4$$

(b) $z = \frac{-11}{3} \Rightarrow x + y = \frac{4}{3}; \ xy = \frac{-55}{3} + 2 = \frac{-49}{3}$

$$x - \frac{49}{3x} = \frac{4}{3} \Rightarrow 3x^2 - 4x - 49 = 0$$

$$x = \frac{2 \pm \sqrt{151}}{3}; \ y = \frac{4}{3} - x = \frac{2 \pm \sqrt{151}}{3}$$

27. $x^2 + xy + xz = 18, y^2 + yz + yx + 12 = 0, z^2 + zx + zy = 30.$

$$x^2 + xy + yz = 18 \Rightarrow x(x + y + z) = 18$$

$$y^2 + yz + yx = -12 \Rightarrow y(x + y + z) = -12$$

$$z^2 + zx + zy = 30 \Rightarrow z(x + y + z) = 30$$

Adding all 3 equations $\Rightarrow (x + y + z)^2 = 36$

$$\Rightarrow x + y + z = \pm 6$$

$$x = \frac{18}{\pm 6} = \pm 3; \quad y = \mp 2; \quad z = \pm 5$$

28. $(y - z)(z + x) = 22, (z + x)(x - y) = 33, (x - y)(y - z) = 6.$

Multiplying $\Rightarrow \left[(x - y)(y - z)(z + x)\right]^2 = 22 \times 33 \times 6$

$$\Rightarrow (x - y)(y - z)(z + x) = \pm 66$$

$$\Rightarrow x - y = \pm 3; \quad y - z = \pm 2; \quad z + x = \pm 11$$

Adding $x - y + y - z + z + x = \pm(3 + 2 + 11) = \pm 16$

$$\Rightarrow x = \pm 8; \quad y = \pm 5; \quad z = \pm 3$$

29. $x^2 y^2 z^2 u = 12, x^2 y^2 z u^2 = 8, x^2 y z^2 u^2 = 1, x y^2 z^2 u^2 = \dfrac{4}{3}.$

Multiplying $\Rightarrow x^7 y^7 z^7 u^7 = 128 \Rightarrow xyzu = 2 \Rightarrow x^2 y^2 z^2 u^2 = 4$

$$\Rightarrow u = \frac{1}{3}; \quad z = \frac{1}{2}; \quad y = 4; x = 3$$

30. $x^3 y^2 z = 12, x^3 y z^3 = 54, x^7 y^3 z^2 = 72$

$$\Rightarrow \frac{z^2}{y} = \frac{54}{12} = \frac{9}{2};$$

$$\frac{x^7 y^3 z^2}{(x^3 y^2 z) \times (x^3 y z^3)} = \frac{x}{z^2} = \frac{72}{12 \times 54} = \frac{1}{9} \Rightarrow 2^2 = 9x$$

$$\Rightarrow \quad z^2 = \frac{9}{2} y = 9x \quad \underset{\text{Or}}{} \quad x = \frac{z^2}{9}; \quad y = \frac{z^2 \times 2}{9}$$

$$\Rightarrow x^3 y^2 z = \frac{z^6}{729} \times \frac{z^4 \times 4}{81} \times z = 12$$

$$\Rightarrow z^{11} = 3^{1+4+6} = 3^{11} \Rightarrow z = 3 \text{ (11 is odd so +3)}$$

$$x = \frac{z^2}{9} = 1; \; y = 2$$

31. $xy + x + y = 23, xz + x + z = 41, yz + y + z = 27.$

$$\Rightarrow x + y + xy + 1 = 24 \Rightarrow (x+1)(y+1) = 24$$

$$(x+1)(z+1) = 42$$

$$(y+1)(z+1) = 28$$

$$\Rightarrow \{(x+1)(y+1)(z+1)\}^2 = 24 \times 42 \times 28$$

$$\Rightarrow (x+1)(y+1)(z+1) = \pm 168$$

$$x + 1 = \pm 6; \; y + 1 = \pm 4; \; z + 1 = \pm 7$$

$$x = 5 \text{ Or} -7; y = 3 \text{or} -5; z = 6 \text{or} -8$$

32. $3x + 8y = 103$

$$\Rightarrow x + \frac{8y}{3} = 34 + \frac{1}{3}$$

$$\Rightarrow x + \frac{8y-1}{3} = 34$$

$$\Rightarrow 3 \text{ divides } 8y - 1$$

$$\frac{16y-2}{3} \text{ is an integer}$$

$$5y + \frac{y-2}{3} \text{ is an integer}$$

Let $\dfrac{y-2}{3} = P$ or $y = 3p + 2$

$$x = \frac{103 - 8y}{3} = \frac{103 - 24p - 16}{3} = 29 - 8_p$$

For positive integers $x \geq 0 \Rightarrow 29 - 8p \geq 0 \Rightarrow p \leq 3...$

$$p = 0, 1, 2, 3$$

$$x = 29, 21, 13, 5$$

$$y = 2, 5, 8, 11$$

33. $5x + 2y = 53$

$$\frac{5}{2}x + y = \frac{53}{2} \Rightarrow y = \frac{53 - 5x}{2} = 26 + \frac{1 - 5x}{2}$$

$$\Rightarrow \frac{5x - 1}{2} \text{ is an integer} \Rightarrow 2x + \frac{x - 1}{2} \text{ is an integer} \Rightarrow \frac{x - 1}{2}$$

is an integer

$$\Rightarrow x = 2p + 1$$

For positive integers, $n > 0 \Rightarrow 53 - 5x > 0 \Rightarrow x \leq 10$

$$p = 0, 1, 2, 3, 4$$

$$x = 1, 3, 5, 7, 9$$

$$y = 24, 19, 14, 9, 4$$

34. $13x + 11y = 414$

$$y = \frac{414 - 13x}{11} = 37 + \frac{7 - 13x}{11}$$

$$\frac{7 - 13x}{11} \text{ is an integer} \Rightarrow \frac{6(7 - 13x)}{11} \text{ is an integer}$$

$$\Rightarrow \frac{42 - 78x}{11} \Rightarrow \frac{9 - x}{11} \text{ is an integer}$$

$$\frac{x - 9}{11} = P \Rightarrow x = 11p + 9$$

$$y \geq 0 \Rightarrow 414 - 13x > 0 \Rightarrow x \leq 31$$

$$p = 0, 1, 2$$

$$x = 9, 20, 31$$

$$y = 27, 14, 1$$

35. $41x + 47y = 2191.$

$$x = \frac{2191 - 47y}{41} = 53 - y + \frac{18 - 6y}{41}$$

$$\Rightarrow \frac{6y - 18}{41} \text{ is an integer} \Rightarrow \frac{42y - 126}{41} \text{ is an integer}$$

$$\Rightarrow \frac{y - 3}{41} = p \text{ Or } y = 41p + 3$$

$$x > 0 \Rightarrow 2191 - 47y > 0 \Rightarrow y \leq 46$$

$$p = 0, 1$$

$$x = 50, 3$$

$$y = 3, 44$$

36. $5x - 7y = 3.$

$$\Rightarrow x = \frac{7y + 3}{5} = y + \frac{2y + 3}{5} \Rightarrow \frac{2y + 3}{5} \text{ is an integer}$$

$$\Rightarrow \frac{6y + 9}{5} \text{ is an integer} \Rightarrow \frac{y + 4}{5} \text{ is an integer} = P$$

$$y = 5p - 4 \Rightarrow x = \frac{35p - 28 + 3}{5} = 7p - 5$$

$$y = 5p - 4; \ x = 7p - 5$$

$$p = 1 \Rightarrow x = 2, y = 1;$$

Smallest positive solution

37. $17y - 13x = 0.$

$$\Rightarrow 17y - 13x = 0 \Rightarrow y = \frac{13x}{17} \text{ Or } x \text{ is divisible by } 17 = 17p$$

$$x = 17p; y = 13p$$

$$p = 1 \Rightarrow x = 17, y = 13 \, ;$$

Smallest positive solution

38. $19y - 23x = 7.$

$$\Rightarrow y = \frac{23x + 7}{19} = x + \frac{4x + 7}{19}$$

$$\Rightarrow \frac{4x + 7}{19} \text{ Is an integer} \Rightarrow \frac{20x + 35}{19} \text{ is an integer}$$

$$\Rightarrow \frac{19x + 38 + x - 3}{19} \Rightarrow \frac{x - 3}{19} \text{ Is an integer} = P$$

$$x = 19p + 3; \quad y = \frac{23(19p + 3) + 7}{19} = 23p + 4$$

$$x = 19p + 3, y = 23p + 4$$

$$p = 0 \Rightarrow x = 3; y = 4 \, ;$$

Smallest positive solution

39. $77y - 30x = 295.$

$$\Rightarrow x = \frac{77y - 295}{30} = 2y - 10 + \frac{17y + 5}{30}$$

$$\Rightarrow \frac{17y + 5}{30} \text{ is an integer} \Rightarrow \frac{7(17y + 15)}{30} = \frac{119y + 35}{30}$$

$$= \frac{120y + 30 + 5 - y}{30} \Rightarrow \frac{5 - y}{30} \text{ is an integer} \Rightarrow \frac{y - 5}{30} \text{ is an}$$

integer = p

$$\Rightarrow y = 30p + 5$$

$$x = \frac{77y - 295}{30} = \frac{77(30p + 5) - 295}{30} = 77p + 3$$

$$p = 0 \Rightarrow x = 3, y = 5 \text{ ; smallest positive integer}$$

40. A farmer spends £752 in buying horses and cows; if each horse costs £37 and each cow £23, how many of each does he buy?

Let him buy h horses and c cows

$$\Rightarrow 37h + 23c = 752; \; h, c > 0$$

$$c = \frac{752 - 374}{23} = 32 - h + \frac{16 - 14h}{23}$$

$$\Rightarrow \frac{16 - 144}{23} \text{ is an integer} \Rightarrow \frac{80 - 70h}{23} \text{ is an integer}$$

$$\Rightarrow \frac{11 - h}{23} \text{ is an integer} \Rightarrow \frac{h - 11}{23} = p \text{ or } h = 23p + 11$$

$$c = \frac{752 - 37(23p + 11)}{23} = 15 - 37p$$

$$p = 0 \Rightarrow h = 11, c = 15;$$

This is the unique solution to the problem. We have 11 horses and 15 cows.

41. Find a number which being divided by 39 gives a remainder 16, and by 56 a remainder 27. How many such numbers are there?

$$x = 39m + 16 = 56n + 27$$

Or $39m + 16 = 56n + 27 \Rightarrow 39m = 56n + 11$

$$m = n + \frac{17n + 11}{39} \Rightarrow \frac{17n + 11}{39} \text{ is an integer}$$

$$\Rightarrow \frac{16(17n + 11)}{39} = \frac{272n + 176}{39} = \frac{273n + 156 + 20 - n}{39} \quad \text{is an}$$

integer

$$\Rightarrow \frac{n-20}{39} = p \text{ Or } n = 39p + 20$$

$$x = 56n + 27 = 2184p + 1147$$

Smallest $x = 1147$ (for p $= 0$)

12 Permutations and Combinations

1. The topic on permutations and combinations deal with the science of selection and science of choice. The typical question we ask in this topic is "in how many ways can we do something" given a constraint.

2. Let us begin our discussion with a few definitions.

3. Permutation: Each of the arrangements which can be made by taking some or all of a number of things is called permutations.

4. Combination: Each of the selection which can be made by taking some or all of a number of things is called a combinations.

5. Arrangements and selections mean very different things. If you had 2 coins - one in red and other in blue.

   (a) We can have 2 arrangements of this coin. Red followed by blue coin; and blue followed by red coin. This can be represented as rb and br .

   (b) How ever we have one only one selection of two coins - one is red and other is blue. Selection does not care about the sequencing of individual components.

   (c) In other words, you can pick the two coins in two ways; but you end up with one selection of red and blue coin at the end.

   (d) This subtle difference between an arrangement and selection is the central notion when we deal with permutations and combinations.

6. The permutations which can be made by taking the letters a, b, c, d two at a time are twelve in number, namely $ab, ac, ad, bc, bd, cd, ba, ca, da, cb, db, dc$. Each of these presenting a different arrangement of two letters.

7. The combinations which can be made by taking the letters a, b, c, d two at a time are six in number namely ab, ac, ad, bc, bd, cd . Each of these presenting a different selection two letters.

8. The previous example clearly outlines the difference between selection and arrangement and hence, the difference between combination and permutation.

9. If one operation can be performed in m ways, and (when it has be performed in any one of these ways) a second operation can then be performed in n ways; the number of ways of performing the two operations will be $m \times n$. If you have 10 buses plying between point A to B, in how many ways can you go from point A to B and return to point A, by a differente bus. You can go from point A to B in 10 ways [pick any one of the 10 buses]; on your way back, pick one of the 9 other buses

10. Let us now consider an example. There are 10 steamers plying between Liverpool and Dublin; in how many ways can a man go from Liver pool to Dublin and return by a different steamer?

    (a) There are ten ways of making the first passage. Why? We have 10 steamers to choose from. If we used one of the steamers, we are left with 9 other steamers for our return journey. Therefore, there are nine ways of returning back. This is because of the constraint of choice "you must return by a different steamer". Hence the number of ways of making the two journeys in $10 \times 9 = 90$.

    (b) This principle may easily be extended to the case in which there are more than two operations each of which can be performed in a given number of ways. Let us consider another example.

11. Three travelers arrive at a town where there are four hotels, in how many ways can they take up their quarters, each at a different hotel?

    (a) The first traveler has choice of four hotels.

    (b) When he has made his selection in any one way, the second traveler has a choice of three hotels

    (c) Similarly, the third traveler can select his hotels in 2 ways

    (d) Hence the required number of ways is $4 \times 3 \times 2, or 24$.

    (e) Therefore, we can conclude that there are 24 ways in which 3 travelers can choose from 4 hotels, each ending up in a different hotel.

12. We have seen the application of permutations in the preceding examples. Let us now take a step further and attempt to generalize our findings.

13. We will now find the number of permutations of n dissimilar things taken r at a time.

   (a) The first place may be filled up in n ways, for any one of the n things may be taken.

   (b) When it has been filled up in any one of these ways, the second place can then be filled up in $(n-1)$ ways;

   (c) Similarly, the third place can be filled up in $(n-2)$ ways

   (d) Quick inferences that can be drawn are:

      i.   ∴ the number of ways in which 3 places can be filled up is $n(n-1)(n-2)$.

      ii. Similarly, the number of ways in which 4 places can be filled up is $n(n-1)(n-2)(n-3)$

      iii. Thus, the number of ways in which r places can be filled up   is $n(n-1)(n-2)...(n-(r-1))$ or $n(n-1)(n-2)...(n-r+1)$

   (e) ∴, The number of ways in which three places can be filled up in $n(n-1)(n-2)$.

   (f) Proceeding in the same way, the number of ways in which r places can be filled up equal to $n(n-1)(n-2)...$ to r factors. And the rth factor is $n-(r-1)=n-r+1$.

   (g) ∴, the number of permutations of n things taken r at a time is $n(n-1)(n-2)...(n-r+1)$.

   (h) Corollary:-

      i.   The number of permutations of n things taken all at a time is $n(n-1)(n-2)...$ to n factors

      ii. Or $n(n-1)(n-2)...3 \cdot 2 \cdot 1$.

      iii. We shall in future denote the number of permutations of n things taken r at a time by the symbol ^nP_r, so that

$$^nP_r = n(n-1)(n-2)...(n-r+1)$$

      iv. $^nP_r = n!$

(i) No concept is complete without looking at an application of the concept. Let us consider an example. Four persons enter a railway carriage with 6 empty seats, in how many ways can they take their places?

    i.   The first person may seat himself in 6 ways

    ii. The second person in 5 ways; the third in 4 ways ; and the fourth in 3 ways

    iii. Since each of these ways may be associated with each of the others, the required answer is $6 \times 5 \times 4 \times 3 = 360$.

(j) Let us now find the number of combinations of n dissimilar things taken r at a time.

Let ^nC_r denote the required number of combinations.

Then each of these combinations consists of a group of r dissimilar things which can be arranged among themselves in $r!$ ways.

Hence $^nC_r \times r!$ is equal to the number of arrangements of n things taken r at a time

In other words, $^nC_r \times r! =\, ^nP_r$

$$= n(n-1)(n-2)...(n-r+1)$$

$$\therefore\, ^nC_r = \frac{n(n-1)(n-2)...(n-r+1)}{r!}$$

$$= \frac{n!}{r! \times (n-r)!}$$

(k) Corollary: $^nC_n = \dfrac{n!}{n! \times (n-n)!} = 1$

(l) Corollary: The number of combinations of n things r at a time is equal to the number of combinations of n things $n-r$ at a time.

We are trying to prove $^nC_r =\ ^nC_{n-r}$

$$^nC_{n-r} = \dfrac{n!}{(n-r)![n-(n-r)]!}$$

$$= \dfrac{n!}{(n-r)!\,r!}$$

$$=\ ^nC_r .$$

As a special case, let us put $r = n$

Then $^nC_0 =\ ^nC_n$.

(m) For example,

From a 15 member team, in how many ways can you choose 11 to play ?

The required number $^{15}C_{11}$.

We can generalize this observation as follows. The number of ways in which $m+n$ things can be divided into two groups containing m and n things is $= \dfrac{(m+n)!}{m!\,n!}$

(n) Let us now find the number of ways in which $m+n+p$ things can be divided into three groups containing m, n, p things.

We will divide and conquer. We will first divide $m+n+p$ things into two groups containing m and $n+p$ things respectively; the number of ways in which this can be done is $\dfrac{(m+n+p)!}{m!(n+p)!}$

Then the number of ways in which the group of $n+p$ things can be divided into two groups containing n and p things respectively is $\dfrac{(n+p)!}{n!p!}$

Hence the number of ways in which the subdivision into three groups containing m, n, p things can be made is

$$\frac{(m+n+p)!}{m!(n+p)!} \times \frac{(n+p)!}{n!p!} = \frac{(m+n+p)!}{m!n!p!}$$

(o) If we consider the case, when m=n=p, we get $\dfrac{(m+n+p)!}{m!m!m!}$.

However, there is a catch here. This result assumes all possible orders in which these groups can occur as well. There are 3! such orders. Therefore the right answer to this question is

$$\frac{(m+n+p)!}{m!m!m!3!}$$

(p) Let us now look at a problem to apply the concept we have looked at.

The number of ways in which 15 recruits can be divided into three equal groups is $\dfrac{15!}{5!5!5!3!}$

The number of ways in which they can be drafted into three different sets of 5 each is $\dfrac{15!}{5!5!5!}$

(q) Let us now find for what value of r, the number of combinations of n things taken r at a time, is greatest.

Since ${}^nC_r = \dfrac{n(n-1)(n-2)...(n-r+2)(n-r+1)}{1\cdot2\cdot3...(r-1)}$

And $^nC_{r-1} = \dfrac{n(n-1)(n-2)...(n-r+2)}{1\cdot2\cdot3...(r-1)}$

$\therefore {}^nC_r = {}^nC_{r-1} \times \dfrac{n-r+1}{r}$

The multiplying factor $\dfrac{n-r+1}{r}$ may be written $\dfrac{n+1}{r}-1$, which shows that it decreases as r increases. Hence as r receives the values $1,2,3...$ in succession, ^nC_r is continually increased until $\dfrac{n+1}{r}-1$ becomes equal to 1 or less than 1.

Now, $\dfrac{n+1}{r}-1>1$

So long as $\dfrac{n+1}{r}>2\dfrac{n+1}{2}>r$

We have to choose the greatest value of r consistent with this in equality

Let n be even, and equal to $2m$, then $\dfrac{n+1}{2}=\dfrac{2m+1}{2}=m+\dfrac{1}{2}$.

And for all values of r up to m inclusive this is greater than r. Hence by putting $r=m=\dfrac{n}{2}$

We find that the greatest number of combinations is $^nC_{\frac{n}{2}}$

Let n be odd and equal to $2m+1$; then $\dfrac{n+1}{2}=\dfrac{2m+2}{2}=m+1$.

And for all values of r up to m inclusive this is greater than r; but when $r=m+1$ the multiplying factor becomes equal to 1

$$^nC_{m+1} =^n C_m \text{ ; that is } ^nC_{\frac{n+1}{2}} =^n C_{\frac{n-1}{2}}$$

And therefore the number of combinations is greatest when the things are taken $\dfrac{n+1}{2} \text{ or } \dfrac{n-1}{2}$ at a time, the result being the same in the two cases.

i. The total number of ways in which it is possible to make a selection by taking some or all out of $p + q + r + \ldots$ things, where of p are of one kind, q are of a second kind, r are of third kind is $(p+1)(q+1)(r+1)\ldots 1$.

12.1   Solved problems

1.   In how many ways can a consonant and a vowel be chosen out of the letters of the word courage?

'Courage' has 4 vowels and 3 Consonants

$$\Rightarrow 4 \times 3 = 12 \text{ Ways of choosing one each}$$

2.   There are 8 candidates for a Classical, 7 for a Mathematical, and 4 for a Natural Science Scholarship. In how many ways can the Scholarships be awarded?

Each scholarship is chosen from the corresponding candidates

$\Rightarrow 8$ ways to choose classical, 7 to choose math and 4 to choose natural science

Totally $8 \times 7 \times 4 = 224$ ways

3.   Find the value of $^8P_7$, $^{25}P_5$, $^{24}C_4$, $^{19}C_{14}$.

$$^8P_7 = 8 \times 7 \times 6 \times 5 \times 4 \times 3 \times 2 \text{ (7 terms)}$$

$$= 40320$$

$$^{25}P_5 = 25 \times 24 \times 23 \times 22 \times 21 \,(5 \text{ terms})$$

$$= 6375600$$

$$^{24}C_4 = \frac{24 \times 23 \times 22 \times 21}{4 \times 3 \times 2 \times 1} = 10626$$

$$^{19}C_{14} = {}^{19}C_5 = \frac{19 \times 18 \times 17 \times 16 \times 15}{5 \times 4 \times 3 \times 2 \times 1} = 11628$$

4. How many different arrangements can be made by taking 5 of the letters of the word equation?

'Equation' has 8 letters

Ans $= {}^8P_5 = 8 \times 7 \times 6 \times 5 \times 4 \,(5 \text{ terms})$

$\qquad = 6720$

5. If four times the number of permutations of n things 3 together is equal to five times the number of permutations of $n-1$ things 3 together, find n.

$\qquad 4 . {}^nP_3 = 5 {}^{n-1}P_3$

$\qquad \Rightarrow 4n(n-1)(n-2) = 5(n-1)(n-2)(n-3)$

$\qquad 4n = 5(n-3)$

$\qquad n = 15$

6. How many permutations can be made out of the letters of the word triangle? How many of these will begin with t and end with e?

'Triangle' has 8 letters

Permutations $= 8! = 40320$

If 't' and 'e' are fixed, remaining 6 letters must be permuted

$\qquad 6! = 720$

7. **Find the number of arrangements that can be made out of the letters of the words (1) independence, (2)superstitious, (3)institutions.**

Independence $= 12$ **letters,** $1i, 3n, 2d, 4e, 1p, 1c$

No of arrangements $= \dfrac{12!}{3!2!4!} = 1663200$

Superstitions $\rightarrow$ **13 letters,** $3s, 2u, 2i, 2t$

No of arrangements $= \dfrac{13!}{3!2!2!2!} = 129729600$

Institutions $\rightarrow$ **12 letters;** $3!, 2u, 2s, 3t,$

No of arrangements $= \dfrac{12!}{3!2!2!3!} = 3326400$

8. **A room is to be decorated with fourteen flags; if 2 of them are blue, 3 red, 2 white, 3 green,2 yellow, and 2 purple, in how many ways can they be hung?**

14 flags, 2 blue, 3 red, 2 white, 3 green, 2 yellow, 2 purple

$$N = \frac{14!}{2!3!2!3!2!2!} = 151351200$$

9. **In how many ways can n things be given to p persons, when there is no restriction as to the number of things each may receive?**

Each things has P choices (persons), n things in all

Total $= p^n$

10. **A library has a copies of one book, b copies of each of two books, c copies of each of three books, and single copies of d books. In how many ways can these books be distributed, if all are out at once?**

No of books $= a + 2b + 3c + d$

$$\text{Ways} = \frac{(a+2b+3c+d)!}{a!b!b!c!c!c!d!} = \frac{(a+2b+3c+d)!}{a!(b!)^2(c!)^3 d!}$$

11. **In how many ways can 7 persons form a ring? In how many ways can 7 Englishmen and 7 Americans sit down at a round table, no two Americans being together?**

Take the first person as an anchor

Remaining 6 can be arranged in 6! Ways

$\Rightarrow$ 6! Different ring arrangements

(b)7 Englishmen can form a ring in 6! Ways sitting in alternate seats. In the empty seats, 7 Americans can be arranged in 7! Ways

Total $= 6! \times 7! = 3628800$ ways

12. Find the sum of all numbers greater than 10000 formed by using the digits 0,2,4,6,8 no digit being repeated in any number.

Since numbers are > 10000; 5 digit numbers

'θ' cannot occur in the 10000 place

Number $= a \times 10^4 + b \times 10^3 + c \times 10^2 + d \times 10 + e$

a is one of 2, 4, 6, 8; occurs as many times as permutations of b, c, d, e $= 4!$

Total $= (2+4+6+8) \times 4! \times 10000 = 4800000$

For b, c, d, e in the other places; they occur they occur as many times as permutations of a and 3 of the rest $= 4! - 3! = 18$ (Since a can't be 0)

Total $= (2+4+6+8+0) \times 18 \times (10+100+1000+1)$

$\qquad = 360 \times 1111 = 399960$

Total $= 5199960$

13 Mathematical Induction

1. We will now discuss a mathematical tool for proof. Many important mathematical formulae are not easily demonstrated by a direct mode of proof; in such cases we frequently find it convenient to employ a method of proof known as Mathematical Induction.

2. How does the Mathematical Induction work as a "proof" tool?

   a. Let us assume that we are asked to prove a equation.

   b. We check the validity of the formula for a small or a trivial value of the variable n.

   c. We then assume that the formula is true for $n = N$.

   d. We then prove that the formula is true for $n = N + 1$.

   e. If this can be done, we can safely conclude that if a formula is valid for $n = N$, it is true for $n = N + 1$.

   f. We have already proved the formula for a trivial value; therefore it must be true for all values of n.

   g. This is the essence of Mathematical Induction.

3. Prove $1^3 + 2^3 + 3^3 \ldots n$ terms $= \dfrac{n(n+1)^2}{2}$

   a. Let us check the validity of the formula for a small value of $n = 1$.

   b. RHS $= \dfrac{1(1+1)}{2^2} = 1 = 1^3 =$ LHS.

   c. $\therefore$, the formula is true for $n = 1$.

   d. Let us assume that the formula is true for $n = n$.

   e. Let us now add the $(n+1)^{th}$ term be added to the LHS and RHS.

   f. This leads us to

   $$1^3 + 2^3 + 3^3 + \ldots + n^3 + (n+1)^3$$

$$= \frac{n(n+1)^2}{2} + (n+1)^3$$

$$= (n+1)^2 \frac{n^2}{4} + (n+1)$$

$$= \frac{(n+1)^2(n^2+4n+4)}{4}$$

$$= \frac{(n+1)^2(n+2)^2}{4}$$

$$= \left[\frac{(n+1)(n+2)}{2} \right]^2$$

g.   Which is of the same form as the result we assumed to be true for n terms, $(n+1)$ taking the place of n

h.   Therefore, the formula is true for all values of n by Mathematical Induction.

4.   Theorems relating to divisibility may often be established by induction.

5.   Let us use the power of Mathematical Induction to show that $x^n - 1$ is divisible by $x - 1$ for all positive integral value of n

a.   By division

$$\frac{x^n - 1}{x - 1} = x^{n-1} + \frac{x^{n-1} - 1}{x - 1}$$

$\therefore x^{n-1} - 1$ is divisible by $x - 1$, then $x^n - 1$ is also divisible by $x - 1$.

But $x^2 - 1, x^3 - 1, x^4 - 1 \ldots$ are divisible by $x - 1$.

Hence the proposition is established.

6.   Mathematical Induction can be applied to all theorems and equations which are applicable across the entire natural number $1, 2, 3 \ldots n$ sequence.

13.1 Solved problems

Prove the following:

1. $1+3+5+.....+(2n-1)=n^2$

$S_1 = 1 = 1^2; \Rightarrow$ true

Assume $S_{n-1} = (n-1)^2$

$S_n = S_{n-1} + t_n = (n-1)^2 + 2n - 1$

$= n^2 - 2n + 1 + 2n - 1 = n^2$

Hence proved

2. $1^2 + 2^2 + 3^2 + ...+ n^2 = \dfrac{1}{6}n(n+1)(2n+1)$

$S_1 = 1\dfrac{1 \times (1+1)(2+1)}{6} = 1; \text{true}$

Assume $S_{n-1} = \dfrac{(n-1)n(2(n-1)+1)}{6} = \dfrac{n(n-1)(2n-1)}{6}$

$S_n = S_{n-1} + t_n = \dfrac{n(n-1)(2n-1)}{6} + n^2$

$= \dfrac{n}{6}\left[(n-1)(2n-1) + 6n\right] = \dfrac{n}{6}\left[2n^2 - 3n + 1 + 6n\right]$

$= \dfrac{n}{6}\left[2n^2 + 3n + 1\right] = \dfrac{n(n+1)(2n+1)}{6}$

Hence proved

3. $2 + 2^2 + 3^2 + ...2^n = 2(2^n - 1)$

$S_1 = 2^1 = 2(2^1 - 1), \text{true}$

Assume $S_{n-1} = 2(2^{n-1} - 1)$

$$S_n = S_{n-1} + t_n = 2^n + 2(2^{n-1} - 1)$$
$$= 2^n + 2 \cdot 2^{n-1} - 2 = 2^n + 2^n - 2 = 2 \cdot 2^n - 2$$
$$= 2(2^n - 1);$$

Hence proved

4. $\dfrac{1}{1.2} + \dfrac{1}{2.3} + \dfrac{1}{3.4} + \ldots$ **to n terms** $= \dfrac{n}{n+1}$

$$S_1 = \frac{1}{1 \cdot 2} = \frac{1}{1+1} = \frac{1}{2} \text{ True}$$

Assume $S_{n-1} = \dfrac{n-1}{n}$

$$S_n = S_{n-1} + t_n = \frac{n-1}{n} + \frac{1}{n(n+1)} = \frac{n(n+1)(n-1) + n}{n^2(n+1)}$$

$$= \frac{n^2 - 1 + 1}{n(n+1)} = \frac{n^2}{n(n+1)} = \frac{n}{n+1},$$

5. Prove by induction that $x^n - y^n$ is divisible by $x + y$ when n is even.

Let $n = 2m$, For $m = 1, 2, 3, \cdots$

For $m = 1, x^2 - y^2 = (x + y)(x - y)$ is divisible

$\qquad m = 2, x^4 - y^4 = (x^2 - y^2)(x^2 + y^2)$ is also divisible

For $m = m - 1$, assume $x^{2m-2} - y^{2m-2}$ is divisible

For $m, x^{2m} - y^{2m}$

$$(x^2 + y^2)\left(x^{2m-2} - y^{2m-2}\right)$$

$$= x^{2m} - y^{2m} - x^2 y^{2m-2} + y^2 x^{2m-2}$$

$$= x^{2m} - y^{2m} + x^2 y^2 \left(x^{2m-4} - y^{2m-1} \right)$$

Or $x^{2m} - y^{2m} = \left(x^{2m-2} - y^{2m-2} \right)\left(x^2 + y^2 \right) - x^2 y^2 \left(x^{2m-4} - y^{2m-4} \right)$

$$T_m = T_{m-1}(x^2 + y^2) - x^2 y^2 T_{m-2}$$

Since T_{m-1} and T_{m-2} are both divisible.

$\therefore T_m$ is also divisible

$my(x + y)$; it follows that so is T_m

14 Binomial Theorem

1. Let us commence our discussion with things that we know from elementary algebra. The student is expected to derive any formula from first principles in case he is not clear about the equations being used.

2. We know that, by actual multiplication that
    $(x+a)\times(x+b)\times(x+c)\times(x+d)$

    $= x^4$

    $+(a+b+c+d)x^3$

    $+(ab+ac+ad+bc+bd+cd)x^2$

    $+(abc+abd+acd+bcd)x$

    $+abcd$

3. Let us now examine the way in which the various partial products are formed, and make a few observations.

(a) The term x^4 is formed by taking the letter x out of each of the factors.

(b) The terms involving x^3 are formed by taking the letter x out of any three factors, in every way possible, and one of the letters a, b, c, d out of the remaining factor.

(c) The terms involving x^2 are formed by taking the letter x out of any two factors, in every way possible, and two of the letters a, b, c, d out of the remaining factors.

(d) The terms involving x are formed by taking the letters x out of any two factor and three of letters a, b, c, d out of the remaining factors.

(e) The term independent of x is the product of all the letters a, b, c, d.

(f) $\therefore$, we can write down the product of the expression $(x-2)$ $(x+3)$ $(x-5)(x+9)$

$$= x^4$$

$$+(-2+3-5+9)x^3$$

$$+(-6+10-18-15+27-45)x^2$$

$$+(30-54+90-135)x$$

$$+270$$

$$= x^4 + 5x^3 - 47x^2 - 69x + 270$$

(g) Similarly, the coefficient of x^3 in the product $(x-3)(x+5)(x-1)(x+2)(x-8)$ can be determined easily.

(h) The terms involving x^3 are formed by multiplying together the x in any three of the factors, and two of the numerical quantities out of the two remaining factors; hence the coefficient is equal to the sum of the products of the quantities $-3, 5, -1, 2, -8$ taken two at a time. Thus the required coefficient is $-15 + 3 - 6 + 24 - 5 + 10 - 40 - 2 + 8 - 16 = -39$.

4. Let us focus our attention on the expansion of $(x+a)^n$ when n is a positive integer.

Let us re-use our conclusions and derivations from the previous section.

$$(x+a)(x+b)(x+c)...(x+k)$$

$$= x^n$$

$$+S_1 x^{n-1}$$

$$+S_2 x^{n-2}$$

$$+...$$

$$+S_r x^{n-r}$$

$$+...$$

$$+S_{n-1}x + S_n$$

In S_1, the number of terms is n

In S_2, the number of terms is the same as the number of combinations of n things 2 at a time; that is n_{C_2}

In S_3 the number of terms is n_{C_3}

... and so on.

Let us now generalize the findings.

i.   If $a = b = c = ... = k$, then

ii.   LHS $= (x + a)^n$, and the coefficients of the expansion on the RHS will be:

iii. $S_1 = {}^n C_1 a$; $S_2 = {}^n C_2 a^2$; $S_3 = nC_3 a^3$

iv.   And $S_k = {}^n C_k a^k$

v. $(x + a)^n = x^n + {}^n C_1 ax^{n-1} + {}^n C_2 a^2 x^{n-2} + ... + {}^n C_n a^n$

vi.   The series containing $n + 1$ terms

vii.   This is the Binomial Theorem, and the expression on the right is said to be the expansion of $(x + a)^n$

6.   The coefficients in the expansion of $(x + a)^n$ are very conveniently expressed by the symbols ${}^n C_1, {}^n C_2, {}^n C_3 ... {}^n C_n$.

We shall, however, sometimes further abbreviate them by omitting n and writing $C_1, C_2, C_3 ... C_n$.

With this notation, we have:

$(x - a)^n$

$= x^n$

$+ C_1 a x^{n-1} + C_2 (-a)^2 x^{n-2}$

$+ C_3 (-a)^3 x^{n-3}$

$+...$

$+C_n(-a)^n$

7. Thus the terms in the expansion of $(x+a)^n$ and $(x-a)^n$ are numerically the same

In $(x-a)^n$ they are alternately positive and negative, and the last term is positive or negative according as n is even or odd.

8. The general terms are the $(r+1)^{th}$ term is given by $T_{r+1} = {}^nC_r x^{n-r}d^r$

Or $T_{r+1} = \dfrac{n(n-1)(n-2)...(n-r+1)}{r!} x^{n-r}a^r$

9. In the general formula for a binomial expansion, if we replace x by 1 and a by x, we get

$$(1+x)^n = 1 + {}^nC_1 x + {}^nC_2 x^2 + ... + {}^nC_r x^r + ... + {}^nC_n x^n$$

$$= 1 + nx + \dfrac{n(n-1)}{1\cdot 2} x^2 + ... + x^n$$

The general term being

$$\dfrac{n(n-1)(n-2)...(n-r+1)}{r!} x^r$$

10. The expansion of a binomial may always be made to depend upon the case in which the first term is unity. Thus

$$(x+y)^n = x(1+\dfrac{y}{x})^n$$

$$= x^n(1+z)^n \text{ where } z = \dfrac{y}{x}.$$

11. Let us now summarize a few observations.

   (a) Binomial Theorem shows symmetry in coefficients. Since
   $${}^nC_k = {}^nC_{n-k}$$

   (b) Again, ${}^nC_0 = {}^nC_n = 1$

(c) Let is now determine the greatest coefficient in the expansion of $(1+x)^n$

(d) The coefficient of the general term of $(1+x)^n$ is ^nC_r and we have only to find for what value of r this is greatest.

(e) When n is even - the greatest coefficient is $^nC_{\frac{n}{2}}$

(f) When n is odd - tt is $^nC_{\frac{n-1}{2}}$ or $^nC_{\frac{n+1}{2}}$; these two coefficients being equal.

12. Let us now look at a way to determine the greatest term in the expansion of $(x+a)^n$

We have, $(x+a)^n = x^n(1+\frac{a}{x})^n$

Since x^n multiplies every term in $(1+\frac{a}{n})^n$, it will be sufficient to find the greatest term in $\left(1+\frac{a}{x}\right)^n$.

Let the rth and $(r+1)^{th}$ terms be any two consecutive terms.

$$(r+1)^{th} \text{ term} = r^{th} \text{ term} \times \frac{n-r+1}{r} \times \frac{a}{x}$$

$$= \left(\frac{n+1}{r}-1\right) \times \frac{a}{x}$$

The factor $\frac{n+1}{r} - 1$ decreases as r increases.

Hence the $(r+1)^{th}$ term is not always greater than the rth term, but only until $(\frac{n+1}{r}-1) \times \frac{a}{x}$ becomes equal to 1, or less than 1.

Now $(\dfrac{n+1}{r} - 1)\dfrac{a}{x} > 1$

So long as $\left(\dfrac{n+1}{r} - 1\right) > \dfrac{x}{a}$

$$\dfrac{n+1}{r} > \dfrac{x}{a} + 1$$

$$\dfrac{n+1}{\dfrac{x}{a}+1} > r$$

If $\dfrac{n+1}{\dfrac{x}{a}+1}$ is an integer P, then if $r = P$ the multiplying factor becomes

1, and the $(p+1)^{th}$ term is equal to the pth, and these terms are greater than any other term.

If $\dfrac{n+1}{\dfrac{x}{a}+1}$ be not an integer, denote its integral part by Q; then the

greatest value of r is Q; hence the $(Q+1)^{th}$ term is the greatest.

13. Let us now find the sum of the coefficients in the expansion of $(1+x)^n$.

$$(1+x)^n = 1 + C_1 x + C_2 x^2 + C_3 x^3 + \ldots + C_n x^n$$

When $x = 1$, all powers of x on the RHS becomes 1; leaving us with the sum of coefficients.

Put $x = 1$,

$$2^n = 1 + C_1 + C_2 + C_3 + \ldots + C_n$$

$$= \text{Sum of the coefficients}$$

Corollary: Therefore, it follows that $C_1 + C_2 + C_3 + \ldots + C_n = 2^n - 1$.

14. Let us now prove that in the expansion of $(1+x)^n$, the sum of the coefficient of the odd terms is equal to the sum of the coefficient of the even terms

$$(1+x)^n = 1 + C_1x + C_2x^2 + \ldots + C_nx^n$$

Put $x = -1$

$$0 = -C_1 + C_2 - C_3 + C_4 - C_5 + \ldots$$

$$1 + C_2 + C_4 + \ldots = C_1 + C_3 + C_5 + \ldots$$

$$\frac{1}{2} \times \text{ (sum of all the coefficients)}$$

$$= \frac{1}{2} \times 2^n$$

$$= 2^{n-1}$$

15. The Binomial Theorem may also be applied to expand expressions which contain more than two terms.

$(x+y+z)^n$ is solved by considering $(y+z) = k$ and solving for $(x+k)^n$.

To this result we substitute the expansion of k^n which is the same as $(y+z)^n$.

14.1 Solved problems

Expand the following:

1.
$$(x^2 + x)^5 = x^5(1+x)^5$$

$$= x^5(1 + 5x + 10x^2 + 10x^3 + 5x^4 + x^5)$$

$$= x^5 + 5x^6 + 10x^7 + 10x^8 + 5x^9 + x^{10}$$

2. $\left[1+\dfrac{x}{2}\right]^{7}$

$$=1+{}^{7}C_{1}\dfrac{x}{2}+{}^{7}C_{2}\dfrac{x^{2}}{4}+{}^{7}C_{3}\dfrac{x^{3}}{8}+{}^{7}C_{4}\dfrac{x^{4}}{16}+{}^{7}C_{5}\dfrac{x^{5}}{32}$$

$$+{}^{7}C_{6}\dfrac{x^{6}}{64}+\dfrac{x^{7}}{128}$$

$$=1+\dfrac{7}{2}x+\dfrac{21}{4}x^{2}+\dfrac{35}{8}x^{3}+\dfrac{35}{16}x^{4}+\dfrac{21}{32}x^{5}+\dfrac{7}{64}x^{6}+\dfrac{x^{7}}{128}$$

3. $\left[\dfrac{2}{3}x-\dfrac{3}{2x}\right]^{6}$

$$=\left(\dfrac{2}{3}x\right)^{6}-{}^{6}C_{1}\left(\dfrac{2}{3}x\right)^{5}\cdot\dfrac{3}{2x}+{}^{6}C_{2}\left(\dfrac{2}{3}x\right)^{4}\left(\dfrac{3}{2x}\right)^{23}$$

$$-{}^{6}C_{3}\left(\dfrac{2}{3}x\right)^{3}\left(\dfrac{3}{2x}\right)^{3}+{}^{6}C_{4}\left(\dfrac{2}{3}x\right)^{2}\left(\dfrac{3}{2x}\right)^{4}$$

$$-{}^{6}C_{5}\left(\dfrac{2}{3}x\right)\left(\dfrac{3}{2x}\right)^{5}+\left(\dfrac{3}{2x}\right)^{6}$$

$$=\dfrac{64}{729}x^{6}-\dfrac{32}{27}x^{4}+\dfrac{20}{3}x^{2}-20+\dfrac{135}{4x^{2}}-\dfrac{243}{8x^{4}}+\dfrac{729}{64x^{6}}$$

4. **5th term of** $\left[2a-\dfrac{b}{3}\right]^{8}$

$$t_{5}={}^{8}C_{4}(2a)^{4}\left(\dfrac{-b}{3}\right)^{4}=\dfrac{{}^{8}C_{4}\times16}{81}a^{4}b^{4}=\dfrac{1120}{81}a^{4}b^{4}$$

5. **7th term of** $\left[\dfrac{4x}{5} - \dfrac{5}{2x}\right]^9$

$$t_7 = {}^9C_6\left(\frac{4x}{5}\right)^3\left(\frac{-5}{2x}\right)^6 = {}^9C_6 \times \frac{4^3}{5^3} \times \frac{5^6}{2^6}\frac{x^3}{x^6}$$

$$= \frac{10500}{x^3}$$

6. **5th term of** $\left[\dfrac{x^{3/2}}{a^{1/2}} - \dfrac{y^{5/2}}{b^{3/2}}\right]^8$

$$t_5 = {}^8C_4\left(\frac{a^{\frac{3}{2}}}{a^{\frac{1}{2}}}\right)^4\left(\frac{-y^{\frac{5}{2}}}{b^{\frac{3}{2}}}\right)^4 = \frac{70x^b\,y^{10}}{a^2b^6}$$

7. $\left(x+\sqrt{2}\right)^4 + \left(x-\sqrt{2}\right)^4$

$$= x^4 + {}^4C_1x^3\sqrt{2} + {}^4C_2x^2\sqrt{2}^2 + {}^4C_3x\sqrt{2}^3 + \sqrt{2}^4$$

$$+ x^4 - {}^4C_1x^3\sqrt{2} + {}^4C_2x^2\sqrt{2}^2 - {}^4C_3x\sqrt{3}^3 + \sqrt{2}^4$$

$$= 2x^4 + 2\cdot 6\cdot 2\cdot x^2 + 2\cdot 4$$

$$= 2x^4 + 24x^2 + 8$$

8. $\left(\sqrt{x^2-a^2}+x\right)^5 - \left(\sqrt{x^2-a^2}-x\right)^5$

Put $\sqrt{x^2-a^2} = y$ or $y^2 = x^2 - a^2$

$$\Rightarrow (y+x)^5 - (y-x)^5$$

$$= \left\{y^5 + {}^5C_1y^4x + {}^5C_2y^3x^2 + {}^5C_3y^2x^3 + {}^5C_4yx^4 + x^5\right\}$$

$$- \left\{y^5 - {}^5C_1y^4x + {}^5C_2y^3x^2 - {}^5C_3y^2x^3 + {}^5C_4yx^4 - x^5\right\}$$

$$= 2 \cdot \left\{ 5y^4 x + 10y^2 x^3 + x^5 \right\}$$

$$= 2 \left\{ 5x(x^2 - a^2) + 10x^3(x^2 - a^2) + x^5 \right\}$$

$$= 2x \left\{ 5(\underline{x^4} - 2a^2 x^2 + a^4) + \underline{10x^4} - 10x^2 a^2 + \underline{x^4} \right\}$$

$$= 2x \left\{ 16x^4 - 20a^2 x^2 + 5a^4 \right\}$$

9. **Find the term independent of x in** $\left[\dfrac{3}{2}x^2 - \dfrac{1}{3x} \right]^9$

We have to determine the term independent of x term in $\left(\dfrac{3x^2}{2} - \dfrac{1}{3x} \right)^9$

General term $= {}^9C_m \left(\dfrac{3}{2}x^2 \right)^{9-m} \left(\dfrac{-1}{3x} \right)^m$

$$= (-1)^m \, {}^9C_m \dfrac{3^{9-2m}}{2^{9-m}} x^{18-3m}$$

Independent of $x \Rightarrow 18 - 3m = 0$ or $m = 6$

Term $= (-1)^6 \, {}^9C_6 \dfrac{3^{-3}}{2^3} = \dfrac{84}{27 \times 8} = \dfrac{7}{18}$

10. **Find the 13th term of** $\left[9x - \dfrac{1}{3\sqrt{x}} \right]^{18}$

$$t_{13} = {}^{18}C_{12}(9x)^6 \left(-\dfrac{1}{3\sqrt{x}} \right)^{12}$$

$$= {}^{18}C_{12} \dfrac{9^6}{3^{12}} \cdot \dfrac{x^6}{\left(\sqrt{x} \right)^{12}} = 18564$$

11. $(x-y)^{30}$ **when** $x = 11, y = 4$

Greatest term of $(x-y)^{30}, x = 11, y = 4$

$$\frac{t_{r+1}}{t_r} = \left(\frac{n+1-r}{r}\right)\frac{a}{x} = \left(\frac{30+1-r}{r}\right)\left(\frac{4}{11}\right)$$

$$= \left(\frac{31-r}{r}\right)\left(\frac{4}{11}\right) \geq 1$$

$$\Rightarrow 124 - 4r \geq 11r \text{ or } 124 \geq 15r$$

$$r \leq 8;$$

Greatest is 9**th** term

$$t_9 = {}^{30}C_8 x^{22} y^8 = \frac{30!}{22!8!} \times 11^{22} 4^8$$

12. $(2x - 3y)^{28}$ **when** $x = 9, y = 4$.

Greatest term of $(2x - 3y)^{28}; x = 9, y = 4$

$$= (2x)^{28}\left(1 - \frac{34}{2x}\right)^{28}$$

$$\frac{T_{r+1}}{T_r} = \frac{28-r+1}{r} \times \frac{3}{2} \times \frac{4}{9} = \frac{29-r}{r} \times \frac{2}{3} = \frac{58-2r}{3r}$$

If $T_{r+1} \geq T_r \Rightarrow 58 - 2r \geq 3r \text{ or } 58 \geq 5r$

$r \leq 11 \Rightarrow$ Greatest term = 12**th**

$$T_{12} = (2x)^{28} \, {}^{28}C_{12}\left(\frac{3y}{2x}\right)^{12}$$

$$= {}^{28}C_{12} \cdot 12^{12} \cdot (18)^{16}$$

$$= \frac{28!}{12!16!} 2^{40} 3^{44}$$

13. $(2a+b)^{14}$ **when** $a = 4, b = 5$

$$= (b+2a)^{14} = b^{14}\left(1 + \frac{2a}{b}\right)^{14}$$

$$\frac{T_{r+1}}{T_r} = \left(\frac{n+1-r}{r}\right)\left(\frac{r}{a}\right) = \frac{15-r}{r} \cdot \frac{1}{\left(\frac{2a}{b}\right)} = \frac{5}{8}\left(\frac{15-r}{r}\right)$$

$$= \frac{75-5r}{8r} \geq 1$$

$$\Rightarrow 75 \geq 13r \text{ Or } r \leq 5$$

Greatest term is the sixth

$$= b^{14}\,^{14}C_5\left(\frac{2a}{b}\right)^5 = 5^{14} \cdot \,^{14}C_5\frac{2^5 \cdot 4^5}{5^5}$$

$$= \frac{14!}{9!5!}5^9 2^{15}$$

14. $(3+2x)^{15}$ **when** $x = \dfrac{5}{2}$

$$= 3^{15}\left(1 + \frac{2x}{3}\right)^{15}$$

$$\frac{T_{r+1}}{T_r} = \frac{n+1-r}{r} \cdot \frac{a}{x} = \frac{16-r}{r} \cdot \frac{2}{3} \cdot \frac{5}{2} = \frac{80-5r}{3r} \geq 1$$

$$\Rightarrow 80 \geq 8r \text{ or } 10 \geq r$$

Greatest term is the 11th; same as the 10th. This is $3^{15}\,^{15}C_{11}(\dfrac{2x}{3})^{11}$

15. Show that the coefficient of the middle term of $(1+x)^{2n}$ is equal to the sum of the coefficients of the middle terms of $(1+x)^{2n-1}$

$(1+x)^{2n}$;

Middle term $= \dfrac{2n+2}{2} = (n+1)^{\text{th}}$

$= {}^{2n}C_n x^n$;

Co-efficient $= {}^{2n}C_n = \dfrac{(2n)!}{n!n!}$

$(1+x)^{2n-1}$

Middle terms are $\dfrac{2n-1+2\pm1}{2} = \dfrac{2n+1\pm1}{2}$

$= n^{\text{th}} \text{and} (n+1)^{\text{th}} \text{terms}$

$T_n = 2n - {}^1C_{n-1}x^{n-1}; T_{n+1} = {}^{2n-1}C_n x^n$

Sum of co-efficient $= \dfrac{(2n-1)!}{(n-1)!n!} + \dfrac{(2n-1)!}{(n-1)!n!}$

$= \dfrac{2\cdot(2n-1)!}{n!(n-1)!} = \dfrac{2n\cdot(2n-1)!}{n\cdot n!(n-1)!} = \dfrac{2n!}{n!n!}$

16. The 2nd, 3rd, 4th terms in the expansion of $(x+y)^n$ are 240, 720, 1080 respectively; find x, y, n.

$r^{\text{th}} \text{ term} = {}^nC_{r-1}x^{n+1-r}y^{r-1}$

2nd term $= {}^nC_1 x^{n-1}y = nx^{n-1}y = 240$

3rd term $= {}^nC_2 x^{n-2}y^3 = \dfrac{n(n-1)}{2}x^{n-2}y^2 = 720$

$$\textbf{4th term} = {}^{n}C_{3}x^{n-2}y^{3} = \frac{n(n-1)(n-2)}{6}x^{n-3}y^{3} = 1080$$

$$\text{Dividing} \frac{\dfrac{n(n-1)}{2}x^{n-2}y^{2}}{nx^{n-1}y} = \frac{720}{240} \Rightarrow \left(\frac{n-1}{2}\right)\frac{y}{x} = 3$$

$$\text{Dividing} \frac{\dfrac{n(n-1)(n-2)}{6}x^{n-3}y^{3}}{\dfrac{n(n-1)}{2}x^{n-2}y^{2}} = \frac{1080}{720} \Rightarrow \left(\frac{n-2}{3}\right)\frac{y}{x} = \frac{3}{2}$$

$$\text{Again dividing} \frac{\left(\dfrac{n-1}{2}\right)\dfrac{y}{x}}{\left(\dfrac{n-2}{3}\right)\dfrac{y}{x}} = \frac{3}{\dfrac{3}{2}} \Rightarrow \frac{3}{2}\left(\frac{n-1}{n-2}\right) = 2$$

Or $\dfrac{n-1}{n-2} = \dfrac{4}{3} \Rightarrow n = 5$

Hence $\dfrac{n-1}{2}\dfrac{y}{x} = 3 \Rightarrow \dfrac{4}{2}\cdot\dfrac{y}{x} = 3 \text{ or } x = \dfrac{2y}{3}$

Substituting; $nx^{n-1}y = 240$

$$\Rightarrow 5\cdot\left(\frac{2y}{3}\right)^{4}\cdot y = 240 \Rightarrow \frac{5\times16\cdot y^{5}}{81} = 240$$

$$y^{5} = \frac{4\times15\times3\times81}{5\times16} = 243 \Rightarrow y = 3$$

$$x = 2$$

Hence $x = 2, y = 3, n = 5$

17. Find the expansion of $(3x^{2} - 2ax + 3a^{2})^{3}$.

$$= \left((3x^{2} - 2ax) + 3a^{2}\right)^{3}$$

$$= (3x^{2} - 2ax)^{3} + 3\cdot(3x^{2} - 2ax)^{2}\cdot3a^{2} + 3(3x^{2} - 2ax)(3a^{2})^{2} + (3a^{2})^{3}$$

$$= \left\{ (3x^2)^3 - 3 \cdot (3x^2)^2(2ax) + 3(3x^2)(2ax)^2 - (2ax)^3 \right\}$$

$$+ 9a^2(9a^4 - 12ax^3 + 4a^2x^2) + 27a^4(3x^2 - 2ax) + 27a^6$$

$$= \underline{27}x^6 - 5\underline{4a}x^5 + 3\underline{6a^2}x^4 - 8\underline{a^3}x^3 + 81\underline{a^2}x^4 - 10\underline{8a^3}x^3 + 36\underline{a^4}x^2$$

$$+ \underline{81}a^4x^4 - 54a^5x + 27a^6$$

$$= 27x^6 - 54ax^5 + 117a^2x^4 - 116a^3x^3 + 117a^4x^2$$

$$- 54a^5x + 57a^6$$

18. **Find the $(p+2)^{th}$ term from the end in $\left[x - \dfrac{1}{x} \right]^{2n+1}$**

An expression with an exponent of $(2n+1)$ will have $(2n+2)$ terms.

Therefore $(p+2)^{th}$ term from the end is the same as

$$= 2n + 1 + 2 - (p+2)$$

$$= 2n + 1 - p \text{ from the start.}$$

We can now apply the equations and formula to solve the problem.

$$t_{2n+1-p} = 2n + {}^1C_{2n-p}x^{p+1}\left(\frac{1}{x} \right)^{2n-p}(-1)^{2n-p}$$

$$\Rightarrow (-1)^{2n-p} = \frac{(1-)^{2n+2p-p}}{(-1)^{2p}} = \frac{(-1)^p}{1} = (-1)^p$$

$$t_{2n+1-p} = \frac{(-1)^p(2n+1)!}{(p+1)!(2n-p)!}x^{2p+1-2n}$$

1: **In the expansion of $(1+x)^{43}$ the coefficients of the $(2r+1)^{th}$ and $(r+2)^{th}$ terms are equal; find r.**

$$t_{2r+1} = {}^{43}C_{2r}x^{2r}; t_{r+2} = {}^{43}C_{r+1}x^{r+1}$$

Co-efficients are equal $\Rightarrow$ $^{43}C_{2r} = \,^{43}C_{r+1}$

$\qquad 2r = r+1 \text{ Or } 2r+r+1 = 43$

$\qquad r = 1 \text{ Or } r = 14$

19. **Find the relation between r and n in order that the coefficients of the $3r^{th}$ and $(r+2)^{th}$ terms of $(1+r)^{2n}$ may be equal.**

$(1+x)^{2n}$; Coefficients of $3r$ and $r+2$ terms are equal

$t_{3r} = \,^{2n}C_{3r-1}x^{3r-1}; t_{r+2} = \,^{2n}C_{r+1}x^{r+1}$

Hence $3r-1 = r+1 \text{ or } (3r-1)+(r+1) = 2n$

$\qquad r = 1 \text{ or } r = \dfrac{n}{2}$

15 Logarithms

1. The logarithm of any number to a given base is the index of the power to which the base must be raised in order to equal the given number.

2. Thus if $a^x = N; x$ is called the logarithm of N to the base a.

3. $3^4 = 81$, logarithm of 81 to base 3 is 4

4. The logarithm of N to base a is usually written as $\log_a N$ and read as log N to the base a.

5. $a^x = N; x = \log_a N$ Also $N = a^{\log_a N}$

6. The logarithm of 1 is 0, $a^0 = 1$

7. $\log 1 = 0$ what ever that base may be

8. The logarithm of the base itself is 1; because $a^1 = a$

9. Let us now find the logarithm of a product.

    (a) Let MN be the product; let a be the base of the system. Thus $x = \log_a M$ and $y = \log_a N$.

    (b) This means $a^x = M, a^y = N$

    (c) Thus the product $MN = a^x a^y = a^{x+y}$

    (d) $\therefore \log_a MN = \log_a M + \log_a N$

    (e) Similarly, $\therefore \log_a ABC = \log_a A + \log_a B + \log_a C$

10. Let us now deteremine of logarithm of a fraction $\dfrac{M}{N}$

    (a) Let $\dfrac{M}{N}$ be the fraction

    (b) And suppose $x = \log_a M; y = \log_a N$

(c) $\therefore \dfrac{M}{N} = \dfrac{a^x}{a^y}$

$= a^{x-y}$

(d) *therefore* $\log_a \dfrac{M}{N} = x - y$

11. Let us now find the logarithm of a number raised to any power, integral or fractional.

(a) $\log_a(M^p)$ has to be determined, given suppose $x = \log_a M$, so that $a^x = M$

(b) The $M^p = (a^x)^p$

$= a^{px}$

$\log_a(M^p) = px$

$\log_a(M^p) = P\log_a M$

(c) $\log_a M^{\frac{1}{r}} = \dfrac{1}{r}\log_a M.$

Common Logarithms

1. Logarithms to the base 10 are called common logarithms

2. From the equation $10^x = N$, it is evident that common logarithms will not in general be integral and that they will not always be positive.

3. For example, $3154 > 10^3$ and $< 10^4$

4. Therefore, $\log 3154 = 3 + a$, where a is a fraction.

5. The integral part of a logarithm is called the characteristic and the decimal part is called the mantissa.

6. To determine the characteristic of the logarithm of any number greater than unity.

7. The characteristic of the logarithm of a number greater than unity is less by one than the number of digits in its integral part, and is positive.

8. The characteristic of the logarithm of a decimal fraction is greater by unity than the number of ciphers immediately after decimal point and is negative.

9. Let N be any number, then since multiplying or dividing by a power of 10 merely alters the position of the decimal point without changing the sequence of figures, it follows that $N \times 10^P$ and $N \div 10^q$, where p and q are any integers, are numbers whose significant digits are the same as those of N.

10. $\log(N \times 10^P) = \log N + P \log 10$

    $= \log N + P$

12. Again, $\log(N \div 10^q) = \log N - \log 10^q$

    $= \log N - q \log 10$

    $= \log N - q$

13. In a logarithm, if an integer is added to log N, and if an integer is subtracted from log N; that is the mantissa or decimal portion of the logarithm remains unaltered.

14. In the case if a negative logarithm the minus sign is written over the characteristic and not before it, to indicate that the characteristic alone is negative, and not the whole expression.

15.1 Solved problems

1. Find the logarithms of 16 to base √2, and 1728 to base 2√3.

(i) Let x be the required logarithm, then by the definition of log

$$\log_{\sqrt{2}} 16 = x$$

Or $(\sqrt{2})^x = 16$

Or $\left[2^{\frac{1}{2}}\right]^x = 2^4$

$$2^{\frac{x}{2}} = 2^4$$

Bases are same, equating powers, we get

$$\frac{x}{2} = 4$$

$$\therefore x = 8$$

(ii) Let $\log_{9\sqrt{3}} 0 \cdot 1 = x$

Or $\left(2\sqrt{3}\right)^x = 1728$

$$\left(2 \; 3^{\frac{1}{2}}\right)^x = 2^6 \cdot 3^3$$

Or $\left(2 \cdot 3^{\frac{1}{2}}\right)^x = 2^6 \cdot \left(3^{\frac{1}{2}}\right)^6$

$$= \left(2 \cdot 3^{\frac{1}{2}}\right)^6$$

Equating powers we get

$$x = 6$$

2. Find the logarithms of 125 to base 5√5 and 0.25 to base 4.

(i) Let $\log_{5\sqrt{5}} 125 = x$

Then $\left(5\sqrt{5}\right)^x = 125$

$$\left(5\sqrt{5}\right)^x = \left(5\sqrt{5}\right)^2$$

$$\therefore x = 2$$

(ii) Let $\log_4 0\cdot25 = x$

Or $4^x = 0\cdot25$

$$4^x = \frac{1}{4}$$

$$4^x = (4)^{-1}$$

$$\therefore x = -1$$

3. **Find the logarithms of** $\dfrac{1}{256}$ **to base** $2\sqrt{2}$ **and** $0\cdot3$ **to base 9.**

Let $\log_{2\sqrt{2}} \dfrac{5}{256} = x$

Then $\left(2\sqrt{2}\right)^x = \dfrac{1}{256}$

$$\left(2^1 \, 2^{\frac{1}{2}}\right)^x = \frac{1}{2^8}$$

$$2^{\frac{3}{2}x} = 2^{-8}$$

Equating powers, we get

$$\frac{3}{2}x = -8$$

$$x = \frac{-16}{3}$$

Also $\log_9 0\cdot3 = x$

Or $(9)^x = 0\cdot3$

$$(3^2)^x = \frac{1}{3}$$

$$3^{2x} = 3^{-1}$$

Or $2x = -1$

$$x = \frac{-1}{2}$$

4. Find the logarithms of 0.0625 to base 2 and 1000 to base 0.01.

(i) Let $\log_2 0 \cdot 0625 = x$

Then $2^x = 0 \cdot 0625$

$$2^x = \frac{1}{16}$$

$$2^x = \frac{1}{2^4}$$

Or $2^x = 2^{-4}$

$\therefore x = -4$

(ii) Also $\log_{0 \cdot 01} 1000 = x$

$$(0 \cdot 01)^x = 1000$$

$$\left(\frac{1}{100}\right)^x = \frac{1}{(1000)^{-1}}$$

Or $\left(\frac{1}{10}\right)^{2x} = \left(\frac{1}{10}\right)^{-3}$

$\therefore 2x = -3$

$$x = \frac{-3}{2}$$

5. Find the logarithms of 0·0001 to base 0·001 and 0·1 to base 9√3.

(i) If x be the required log x, then

$$\log_{0 \cdot 001} 0 \cdot 0001 = x$$

Or $(0 \cdot 001)^x = 0 \cdot 0001.$

$$\left(\frac{1}{10^3}\right)^x = \frac{1}{10^4}.$$

$$\left(\frac{1}{10}\right)^{3x} = \left(\frac{1}{10}\right)^4$$

Or $3x = 4$

$$x = \frac{4}{3}$$

(ii) Let $\log_{9\sqrt{3}} 0 \cdot 1 = x$

Or $\left(9\sqrt{3}\right)^x = 0 \cdot 1$

$$\left(3^2 \cdot 3^{\frac{1}{2}}\right)^x = \frac{1}{9}.$$

$$3^{\frac{5}{2}x} = 3^{-2}$$

Or $\frac{5}{2}x = -2$

$$x = \frac{-4}{5}$$

6. **Find by inspection the characteristics of the logarithms of 21735, 23·8, 350, ·035, 0.2, 0.87, 0.875**

Characteristic of logarithm 21735 will be 4

(∵ No digits is 5, subtracting 1 from the number of digits).

Characteristic of logarithm 23.8 will be 1.

Characteristic of logarithm 350 will be 2.

Characteristic of logarithm 0.035 will be 2.

(Adding one to the number of ciphers immediately after decimal)

Characteristic of logarithm 0.2 will be 1.

Characteristic of logarithm 0.87 will be 1.

Characteristic of logarithm 0.875 will be .1

7. The mantissa "a" of log .7623 is 0.9921259 write down the logarithms of 7623, 7623, 0.007623, 762300, 0.000007623.

Given the mantissa of $7623 = 0.8821259$.

$\therefore \log 7.623 = 0.8821259.$

$\log 762.3 = 2.8821259$

$\log 0 \cdot 007623 = \overline{3} \cdot 8821259$

$\log 762300 = 5 \cdot 8821259$

$\log 0 \cdot 000007623 = \overline{6} \cdot 8821259$

8. Find by inspection the characteristics of logarithms of : 21735, 23.8, 350, .035, .2, .875.

The numbers of digits in the integral parts are 5, 2, 4 and 6 respectively.

9. Find the value of log 64, given log 2 = 0.3010300.

$\log 64 = \log 2^6$

$= 6 \log 2$

$= 6(0.30110300)$

$= 1.8061800.$

16 Inequalities

1. Any quantity a is said to be greater than another quantity when $a - b$ is positive.

2. b is said to be less than a when $b - a$ is negative.

   (a) 2 is greater than -3, because $2 - (-3)$ is positive.

   (b) -5 is less than -2, because $-5 - (-2) = -3(-ve)$

3. If $a > b$, and if $c > 0$, then it is evident that

   (a) $a + c > b + c$

   (b) $a - c > b - c$

   (c) $ac > bc$

   (d) $\dfrac{a}{c} > \dfrac{b}{c}$

   (e) That is, an inequality will still hold after each side has been increased diminished, multiplied, or divided by the same positive quantity.

4. If $a - c > b$ then $a > b + c$

   This shows that in an inequality any term may be transposed from one side to the other if its sign be changed

5. If $a > b$ then evidently $b < a$

   This means if the sides of an inequality be transposed, the sign of inequality must be reversed.

6. If $a > b$ then $a - b$ is positive and $b - a$ is negative

   This means $-a - (-b)$ is negative, and therefore $-a < -b$;

7. Hence, if the signs of all the terms of an inequality be changed, the sign of inequality must be reversed.

   Again, if $a > b$, then $-a < -b$, and therefore $-ac < -bc$

   This implies that if the sides of an inequality be multiplied by the same negative quantity, the sign of inequality must be reversed.

8. If $a_1 > b_1, a_2 > b_2, a_3 > b_3 \ldots a_3 > b_3$, it is clear that
    $$a_1 + a_2 + a_3 + \ldots + a_m > b_1 + b_2 + b_3 + \ldots + b_m$$

    And $a_1 a_2 a_3 \ldots a_m > b_1 b_2 b_3 \ldots b_m$.

9. If $a > b$, and if p, q are positive integers, then

    $$\sqrt[q]{a} > \sqrt[q]{b}$$

    $$a^{\frac{1}{q}} > b^{\frac{1}{q}}$$

    $$a^{\frac{p}{q}} > b^{\frac{p}{q}}$$

    $a^n > b^n$, where n is any positive quantity

    $$\frac{1}{a^n} < \frac{1}{b^n}, a^{-n} < b^{-n}.$$

10. The square of every real quantity is positive, and therefore greater than zero.

    $(a - b)^2 > 0$; and therefore positive.

    $$a^2 - 2ab + b^2 > 0$$
    $$a^2 + b^2 > 2ab$$

11. The arithmetic mean of two positive quantities is greater than their geometric mean. This can be expressed as $\dfrac{x + y}{2} > \sqrt{xy}$

12. Let a and b, be two positive quantities, S be their sum and P their product; we can deduce the following.

    $$S = a + b$$
    $$P = ab$$
    $$4ab = (a + b)^2 - (a - b)^2$$
    $$4P = S^2 - (a - b)^2 \text{ and } S^2 = 4P + (a - b)^2$$

Hence, if S is given P is greatest when $a = b$; and if P is given, S is least when $a = b$;

If the sum of two positive quantities is given, their product is greatest when they are equal; and of the product of two positive quantities is given, their sum is least when they are equal.

13. The arithmetic mean of any number of positive quantities is greater than the geometric mean.

14. Let us now determine the greatest value of $a^m b^n c^p \ldots$ when $a + b + c + \ldots$ is constant; m, n, p being positive integers.

Since $m, n, p \ldots$ are constants, the expression $a^m b^n c^p \ldots$ will be greatest when $\dfrac{a^m}{m} \dfrac{b^n}{n} \dfrac{c^p}{p} \ldots$ is greatest.

This expression is the product of $m + n + p + \ldots$ factors whose sum is

$$m\frac{a}{m} + n\frac{b}{n} + p\frac{c}{p} + \ldots, \text{ or } a + b + c + \ldots \text{ and therefore constant.}$$

Hence $a^m b^n c^p \ldots$ will be greatest when the factors $\dfrac{a}{m}, \dfrac{b}{n}, \dfrac{c}{p} \ldots$ are all equal.

$$\frac{a}{m} = \frac{b}{n} = \frac{c}{p} = \ldots = \frac{a+b+c+\ldots}{m+n+p+\ldots}$$

Thus the greatest value is

$$m^n n^n p^p \ldots (\frac{a+b+c+\ldots}{m+n+p+\ldots})^{m+n+p+\ldots}$$

15. If there are n positive quantities $a, b, c \ldots K$, then
$$\frac{a^m + b^m + c^m + \ldots + k^m}{n} > (\frac{a+b+c+\cdots+k}{n})^m, \text{ unless m is a}$$
positive proper fraction

16. If a and b are positive integers, and $a > b$ and if x be a positive quantity $(1 + \frac{x}{a})^a > (1 + \frac{x}{b})^b$

$(1 + \frac{x}{a})^a = 1 + x + (1 - \frac{1}{a})\frac{x^2}{2!} + (1 - \frac{1}{a})(1 - \frac{2}{a})\frac{x^3}{3!} + \ldots$ This series consists of $a + 1$ terms.

$(1 + \frac{x}{a})^b = 1 + x + (1 - \frac{1}{b})\frac{x^2}{2!} + (1 - \frac{1}{n})(1 - \frac{2}{b})\frac{x^3}{3!} + \ldots$ This series consists of $b + 1$ terms.

After the second term, each term in the first equation is greater than the corresponding term in the second equation.

The number of terms in the first equation is greater than the number of terms in the first equation.

This proves the validity of the proposition.

16.1 Solved problems

1. **Prove that** $(ab + xy)(ax + by) > 4abxy$.

Since AM > GM, therefore

$$\frac{ab + xy}{2} > (ab \cdot xy)^{\frac{1}{2}}$$

Or $(ab + xy) > 2(ab \cdot xy)^{\frac{1}{2}}$ (1)

Similarly

$$\left(\frac{ax + by}{2}\right) > (ax \cdot by)^{\frac{1}{2}}$$

$$ax + by > 2(abxy)^{\frac{1}{2}} \quad (2)$$

By multiplying (1) & (2), we get

$$(ab + xy)(ax + by) > 2(ab \cdot xy)^{\frac{1}{2}}(ab \cdot xy)^{\frac{1}{2}}$$

Or $(ab + xy)(ax + by) > 4abxy$

2. Prove that $(b + c)(c + a)(a + b) > 8abc$.

We know that Arithmetic mean of any number of quantities is greater than their geometric means.

$$\frac{b+c}{2} > (bc)^{\frac{1}{2}} \text{ Or } (b+c) > 2(bc)^{\frac{1}{2}} \text{ (1)}$$

Similarly $\dfrac{c+a}{2} > (ca)^{\frac{1}{2}}$ or $(c+a) > 2(ca)^{\frac{1}{2}}$ (2)

And $\dfrac{a+b}{2} > (ab)^{\frac{1}{2}}$ or $(a+b) > (ab)^{\frac{1}{2}}$ (3)

By multiplying (1),(2), and (3), we get

$$(a+b)(b+c)(c+a) > 2(ab)^{\frac{1}{2}} \cdot 2(bc)^{\frac{1}{2}} \cdot 2(ca)^{\frac{1}{2}}$$

$$> 8(a^2 b^2 c^2)^{\frac{1}{2}}$$

$$\therefore (a+b)(b+c)(c+a) > 8abc$$

3. Show that the sum of any real positive quantity and its reciprocal is never less than two.

Suppose that any real positive quantity is a. Then its reciprocals is $\dfrac{1}{a}$.

We have to prove that $a + \dfrac{1}{a} > 0$

Let us assume that $a + \dfrac{1}{a} < 2$

Or $a + \dfrac{1}{a} - 2 < 0$

Or $\left(\sqrt{a}\right)^2 + \left(\dfrac{1}{\sqrt{a}}\right)^2 - 2\sqrt{a} \times \dfrac{1}{\sqrt{a}} < 0$

Or $\left(\sqrt{a} - \dfrac{1}{\sqrt{a}}\right)^2 < 0$; which is a contradiction.

Since the square of any real quantity cannot be less than zero, hence the result is greater than zero.

4. **If** $a^2 + b^2 = 1$ **and** $x^2 + y^2 = 1$ **show that** $ax + by < 1$.

Given that $a^2 + b^2 = 1$ and $x^2 + y^2 = 1$

We know that arithmetic mean by two quantities is greater than their geometric mean.

$$\therefore \frac{a^2 + x^2}{2} > (a^2 x^2)^{\frac{1}{2}} \text{ Or } a^2 + x^2 > 2ax \quad (1)$$

Similarly

$$\frac{b^2 + y^2}{2} > (b^2 y^2)^{\frac{1}{2}} \text{ or } b^2 + y^2 > 2by \quad (2)$$

Adding (1) & (2), we get

$$a^2 + x^2 + b^2 + y^2 > 2ax + 2by \; (\because a^2 + b^2 = 1$$

$$a^2 + b^2 + x^2 + y^2 > 2(ax + by) \; x^2 + y^2 = 1$$

$$1 + 1 > 2(ax + by)$$

Or $2 > 2(ax + by)$

$$1 > ax + by$$

Or $ax + by < 1$

5. **If** $a^2 + b^2 + c^2 = 1,$ **and** $x^2 + y^2 + z^2 = 1,$ **show that** $ax + by + cz < 1.$

We know that arithmetic mean of two quantities is greater than their geometric mean.

$$\therefore \frac{a^2 + x^2}{2} > (a^2 x^2)^{1/2}$$

Or $a^2 + x^2 > 2ax$ (1)

Similarly $\dfrac{b^2 + y^2}{2} > (b^2 y^2)^{1/2}$

Or $b^2 + y^2 > 2by$ (2)

And $\dfrac{c^2 + z^2}{2} > (c^2 z^2)^{1/2}$

Or $c^2 + z^2 > 2cz$ (3)

Hence by adding (1), (2) and (3), we get

$$a^2 + x^2 + b^2 + y^2 + c^2 + z^2 > 2ax + 2by + 2cz$$

Or $a^2 + b^2 + c^2 + x^2 + y^2 + z^2 > 2(ax + by + cz)$

$$1 + 1 > 2(ax + by + cz) \; (\because a^2 + b^2 + c^2 = 1$$

$$2 > 2(ax + by + cz) \; x^2 + y^2 + z^2 = 1.$$

Or $1 > ax + by + cz$

Or $ax + by + cz < 1$

6. **If** $a > b$, **show that** $a^a b^b > a^b, b^a$ **and** $\left(\dfrac{b}{a}\right) < \log\left(\dfrac{1+b}{1+a}\right)$

Since $a > b$.

$$\therefore a - b > 0$$

i.e. $a - b$ is positive.

$$\therefore a^{a-b} > b^{a-b}$$

Or $\dfrac{a^a}{a^b} > \dfrac{b^a}{b^b}$

Or by cross multiplication

$$a^a b^b > a^b b^a$$

Hence the result

Again $a > b$ or $b < a$

On adding ab on both sides, we get

$$b + ab < a + ab$$

Or $b(1 + a) < a(1 + b)$

Taking log both sides, we get

$$\log b + \log(1 + a) < \log a + \log(1 + b)$$

Or $\log b - \log a < \log(1 + b) - \log(1 + a)$

Or $\log\left(\dfrac{b}{a}\right) < \log\left(\dfrac{1 + b}{1 + a}\right)$

Hence the result

7. Show that $(x^2 y + y^2 z + z^2 x)(xy^2 + yx^2 + zx^2) > 9x^2 y^2 z^2$.

We know that Arithmetic mean of any no of quantities is greater than their geometric mean.

$$\therefore \quad \frac{x^2 y + y^2 z + z^2 x}{3} > (x^2 y \cdot y^2 z \cdot z^2 x)^{\frac{1}{3}}$$

$$> (x^3 y^3 z^3)^{\frac{1}{3}}$$

$$> x y z$$

Or $x^2 y + y^2 z + z^2 x > 3xyz$ (1)

Similarly $\dfrac{xy^2 + yz^2 + zx^2}{3} > (xy^2 \cdot yz^2 \cdot zx^2)^{\frac{1}{3}}$

$$xy^2 + yz^2 + zx^2 > 3(x^3 y^3 z^3)^{\frac{1}{3}}$$

$$xy^2 + yz^2 + zx^2 > 3xyz \quad (2)$$

On multiplying (1) and (2), we get

$$(x^2 y + y^2 z + z^2 x)(xy^2 + yz^2 + zx^2) > (3xyz)(3xyz)$$

$$\therefore (x^2 y + y^2 z + z^2 x)(xy^2 + yz^2 + zx^2) > 9x^2 y^2 z^2$$

8.   Find which is the greater $3ab^2$ **or** $a^3 + 2b^3$.

Since $a^3 + 2b^3 - 3ab^2 = a^3 - ab^2 + 2b^3 - 2ab^2$

$$= a(a^2 - b^2) + 2b^2(b - a)$$

$$= a(a + b)(a - b) - 2b^2(a - b)$$

$$= (a - b)[a^2 + ab - 2b^2]$$

But $a^2 + ab - 2b^2 = a^2 - b^2 + ab - b^2$

$$= (a + b)(a - b) + b(a - b)$$

$$= (a - b)(a + b + b)$$

$$= (a - b)(a + 2b)$$

$$\therefore a^3 + 2b^3 - 3ab^2 = (a - b)(a - b)(a + 2b)$$

$$= (a - b)^2 (a + 2b)$$

Since left hand side is always positive.

i.e. greater than zero.

i.e. $a^3 + 2b^3 > 3ab^2$

Hence $a^3 + 2b^3$ is the greater.

9.   Prove that $a^3 b + ab^3 < a^4 + b^4$.

$$\because a^4 + b^4 - a^3 b - ab^3 = a^4 - a^3 b + b^4 - ab^3$$

$$= a^3(a - b) + b^3(b - a)$$

$$= (a - b)[a^3 - b^3]$$

$$= (a - b)(a - b)(a^2 + ab + b^2)$$

$$= (a - b)^2 (a^2 + ab + b^2)$$

This is always positive.

$$\therefore a^4 + b^4 - a^3 b - ab^3 > 0$$

Or $a^4 + b^4 > a^3 b + ab^3$

i.e. $a^3 b + ab^3 < a^4 + b^4$

Hence the result

10. Prove that $6abc < bc(b+c) + ca(c+a) + ab(a+b)$.

Or $6abc < b^2 c + bc^2 + c^2 a + ca^2 + a^2 b + ab^2$

Or $6abc < a(b^2 + c^2) + b(c^2 + a^2) + c(a^2 + b^2)$

Or $a(b^2 + c^2) + b(c^2 + a^2) + c(a^2 + b^2) > 0$

We know that arithmetic mean of two quantities is greater than their geometric mean.

$$\therefore \frac{b^2 + c^2}{2} > (b^2 c^2)^{1/2} = bc$$

Or $b^2 + c^2 > 2bc$.

And hence on multiplying by a on both sides, we get

$$a(b^2 + c^2) > 2abc. \quad (1)$$

Similarly

$$\frac{c^2 + a^2}{2} > (c^2 a^2)^{1/2} = ca$$

Or $b(c^2 + a^2) > 2abc. \quad (2)$

And $\dfrac{a^2 + b^2}{2} > (a^2 b^2)^{1/2} = ab$

Or $c(a^2 + b^2) > 2abc \quad (3)$

On adding (1), (2) and (3), we get

$$a(b^2 + c^2) + b(c^2 + a^2) + c(a^2 + b^2) > 2abc + 2abc + 2abc$$

Or $a(b^2 + c^2) + b(c^2 + a^2) + c(a^2 + b^2) > 6abc$.

17 Probabilities

1. Probability deals with the study of chance.
2. If an event can happen in a ways and fail in b ways, and each of these ways is equally likely, the probability, or the chance, of its happening is $\dfrac{a}{a+b}$, and that of its failing is $\dfrac{b}{a+b}$.
3. If P is the probability of the happening of an event, the probability of its not happening is $1-p$.
4. Let us consider the tossing of a coin. The coin has two faces or sides---the heads and the tails. If the coin is unbiased and fair, if we keep tossing the coin, one of the two faces will appear randomly. In this case, we can see the following:
    a. Number of possible outcomes = Heads or Tails. Therefore, we have 2 possible outcomes.
    b. Probability of Heads $=\dfrac{1}{2}$
    c. Probability of Tails $=\dfrac{1}{2}$
5. What is the chance of throwing a number greater than 3 with an ordinary die whose faces are numbered from 1 to 6?
    a. Number of possible outcomes = 6; because there are 6 possible ways in which the die can fall.
    b. Number of favorable outcomes = 3, because we have three of them---4,5,6
    c. Probability of throwing a number greater than three is $=\dfrac{3}{6}=\dfrac{1}{2}$
    d. Probability of NOT throwing a number greater than three is $=1-\dfrac{1}{2}=\dfrac{1}{2}$
6. General strategy for solving problems simply involves the following 4 steps.
    a. Step 1: Determine the total number of outcomes of the event: E

b. Step 2: Determine the total number of FAVORABLE outcomes : F

c. Step 3: Required Probability of the favorable outcome
: $P = \dfrac{F}{E}$

d. Step 4: Remember that $F \leq E, \therefore P \leq 1$. This may be used as a check for correctness.

7. Suppose that there are a number of events A, B, C,... of which one must, and only one can, occur also suppose that a, b, c,.... The numbers of ways in which these events can happen may be determined by finding out the chance of each of the events.

a. Total number of outcomes : $a + b + c + \cdots$

b. The number of outcomes favorable to $A = a$

c. $\therefore$, the chance that A will happen $= \dfrac{a}{a + b + c + \cdots}$

d. Similarly the chance that B will happen is
$\dfrac{b}{a + b + c + \cdots}$

8. A compound event simply represents the occurence of two or more connected events. In other words, this deals with probabilities when two or more possibilities occur in connection with each other.

a. To throw more light on this, let us consider the following example. For example suppose we have a bag containing 6 white socks and 8 black socks. Let us assume that we wish to draw 3 socks at a time. If we wish to estimate the chance of drawing first 3 white socks and followed by 3 black socks.

b. Let us consider this event carefully. We start off with 14 socks in all---6 white and 8 black ones.

c. After the first drawing, we are left with 11 socks in the bag. Therefore, our seccond draw is an event of picking 3 socks from 11 left in the bag.

d. This is an example of a compound event.

9. Events are said to be dependent or independent according as the occurrence of the others. Dependent events are sometimes said to be contingent.

    a.   If we tossed two unbiased coins, the outcome of one coin is independent of the outcome of the other. These are independent events.

    b.   A compound event is a good example of a dependent event.

10. We just looked a few definitions and scenarios. We will now turn our attention to computing probabilities. If there are two independent events the respective probabilities of which are known. Our task is to find the probability that both will happen.

    a.   Suppose that the first event may happen in a ways and fail in b ways, all these cases being equally likely; and suppose that the second event may happen in a_1 ways and fail in b_1 ways, all these ways being equally likely.

    b.   Probability that both events happen

    c. $= \dfrac{aa_1}{(a+b)(a_1+b_1)}$

    d.   Probability that both events fail to happen

    e. $= \dfrac{bb_1}{(a+b)(a_1+b_1)}$

    f.   Probability that the first event happens and the second event fails

    g. $= \dfrac{ab_1}{(a+b)(a_1+b_1)}$

    h.   Probability that the first event fails and the second event happens

    i. $= \dfrac{a_1b}{(a+b)(a_1+b_1)}$

11. Let us assume that $p_1, p_2 p_3 \cdots$ respectively, are the probabilities that a number of independent events will separately happen. Then,

    a.   The chance that they will all happen $= p_1 \times p_2 \times p_3 \cdots$

    b.   The chance that the two first will happen and the rest fail $= p_1 \times p_2 \times (1-p_3) \times (1-p_4) \cdots$

    c.   If p is the chance that an event will happen in one trial. The chance that will happen in any assigned succession of r trials $= p^r$.

    d.   Suppose $p_1 = p_2 = p_3 = \cdots = p$

    e.   To find the chance that at least one of the events will happen is determined as follows.

    f.   The chance that all the events fail is
$$(1 - p_1)(1 - p_2)(1 - p_3) \cdots$$

    g.   Hence the probability that atleast one of the event happen
$$= 1 - (1 - p_1)(1 - p_2)(1 - p_3) \cdots$$

12. If an event can happen in two or more different ways which are mutually exclusive, the chance that it will happen is the sum of the chances of its happening in these different ways.

13. If an event can happen in n ways which are mutually exclusive, and if $p_1, p_2, p_3, \cdots p_n$ are the probabilities that the event will happen in these different ways respectively, the probability that it will happen in some one of these ways is $p_1 + p_2 + p_3 + \cdots + p_n$.

14. The probability of one or other of a series of events is the sum of the probabilities of the separate events only when the events are mutually exclusive, that is when the occurrence of one is in compatible with the occurrence of any of the others.

15. The probability of the happening of an event in one trial being known, required the probability of its happening once, twice, thrice times, exactly in n trials.

    a.   Let P be the probability of the happening of the event in a single trial, and let $q = 1 - p$; then the probability that the event will happen exactly r times in n trials is the $(r+1)$th term in the expansion of $(q + p)^n$.

    b.   If we expand $(p + q)^n$ by the Binomial Theorem, we have

    c.
$$p^n +\,^n C_1 p^{n-1} q +\,^n C_2 p^{n-2} q^2 + \cdots$$
$$+\,^n C_{n-r} p^r q^{n-r} + \cdots + q^n$$

    d. Thus the terms of this series will represent respectively the probabilities of the happening of the event exactly n times, $n-1$ times, $n-2$ times, ... in n trials.

    e. If P represents a person's chances of success in any venture and M the sum of money which he will receive in case of success, the sum of money denoted by PM is called his expectation.

16. If P is the probability that an event happens in a single trial, then if the number of trials is indefinitely increased, it becomes a certainly that the limit of the ratio of the number of successes to the number of trials is equal to P in other words, if the number of trials is N, the number of successes may be taken to be PN.

17. Let us consider the following. An observed event has happened through some one of a number of mutually exclusive causes. We are required to find the probability of any assigned cause being the true one.

    a. Let there be n causes, and before the event took place suppose that the probability of the existence of these causes was estimated at $p_1, p_2, p_3, \cdots p_n$.

    b. Let p_r denote the probability that when the rth cause exists the event will follow: after the event has occurred it is required to find the probability that the rth cause was the true one.

    c. Consider N trials, where N is a very large number. Then the first cause exists in $p_1 N$ of these, and out of this number the event follows in $p_1 P_1 N$.

    d. Similarly, there are $p_2 P_2 N$ trials in which the event follows from the second cause: and soon for each of the other causes. Hence the number of trials in which the event follows is $(p_1 P_1 + p_2 P_2 + \cdots + p_n P_n)N$, or

$$N\sum(pP)$$

    e. And the number in which the event was due to the rth cause is $p_r P_r N$.

f. Hence after the event the probability that the rth cause was the true one is $p_r P_r N \div N \sum (pP)$.

g. That is, the probability that the event was produced by the rth cause is $\dfrac{p_r P_r}{\sum pP}$.

18. Let us turn our attention to another situation. Here, we consider an event that has happened through any one of the several mutually exclusive causes; and we are required to find the probability of any assigned cause being the true one.

  a. Let there be n causes, and before the event took place suppose that the probability of the existence of these causes was estimated at $p_1, p_2, p_3, \cdots p_n$.

  b. Let p_r denote the probability that when the rth cause exists the event will follow; then the antecedent probability that the event would follow from the rth causes is $p_r P_r$.

  c. Let Q_r be the posteriori probability that the rth cause was the true one; then the probability that the rth cause was the true one is proportional to the probability that if in existence, this cause would produce the event.

  d. ∴
$$\frac{Q_1}{p_1 P_1} =$$
$$= \frac{Q_2}{p_2 P_2}$$
$$= \cdots$$
$$= \frac{Q_n}{p_n P_n}$$
$$= \frac{\sum Q}{\sum pP}$$

$$= \frac{1}{\sum pP}$$

e. $Q_r = \frac{p_r P_r}{\sum pP}.$

17.1 Solved problems

1. **In a single throw with two dice, find the chances of throwing a sum of (1) five, (2) six.**

Two dice may be thrown in 6×6 . ie. Ways.

(1). And the cases favorable to the event that the sum on two dice is five are $(1,4)(2,3)(3,2) \& (4,1)$.

i.e. 4 cases.

Hence chances of throwing two dice in such a way that total on two dice is five $= \frac{4}{36} = \frac{1}{9}.$

(2). The cases favorable to the event that the sum on two dice is six are $(1,5)(2,4)(3,3)(4,2)$ and $(5,1)$ i.e. 5

Hence the chances of throwing six on two dice is $= \frac{5}{36}.$

2. **From a pack of 52 cards, two are drawn at random; find the chance that one is a Knave and the other a Queen.**

Given a pack contains 52 cards.

From a pack of 52 cards, two are drawn at random in $^{52}C_2$ ways

$$= \frac{52!}{(52-2)! \, 2!}$$

$$= \frac{52 \times 51 \times \cancel{50!}}{\cancel{50!} \, 2!}$$

$$= 26 \times 51$$

$$= 1326.$$

The cases favorable to the event that one is a knave and the other a queen when drawn together are

$$^4C_1 \times {}^4C_1 = \frac{4!}{3! \times 1!} \times \frac{4!}{3! \times 1!}$$

$$= \frac{4 \times 3!}{3!} \times \frac{4 \times 3!}{3!}$$

$$= 4 \times 4$$

$$= 16.$$

In a pack of 52 cards since there are 4 knaves and 4 queens. Out of 4 knaves, one knaves can be drawn in $^4C_1$ ways and similarly, one queen can be drawn in $^4C_1$ and each of the event can be combined together and hence the favorable cases are $^4C_1 \times {}^4C_1 = 16$.

Hence the required probability $= \dfrac{16}{1326} = \dfrac{8}{663}$

3. A bag contains 5 white, 7 black and 4 red balls, find the chance that three balls drawn at random are all white.

Total no of ball = 5white + 7black + 4red balls

$$= 16\text{balls}.$$

Out of 16 balls 3 can drawn in $^{16}C_3$ ways

$$= \frac{16!}{(16-3)!3!}$$

$$= \frac{16 \times 15 \times 14 \times 13!}{13! \times 6.}$$

$$= \frac{16 \times 15 \times 14}{6} = 8 \times 5 \times 14$$

Total white balls are 5.

The cases favorable to the event that the balls drawn are white are

$$^5C_3 = \frac{5!}{2!\,3!}$$

$$= 10.$$

Hence the required probability $= \dfrac{\cancel{10}^{2}}{8 \times 5 \times \cancel{14}_{\,7}}$

$$= \frac{1}{56}$$

4. If four coins are tossed, find the chance that there should be two heads and two tails

Let the outcome of tossing a coin when it turns up head or tail be denoted by H or T respectively.

Now the numbers of ways of tossing 4 coins are $2 \times 2 \times 2 \times 2 = 16$.

Again,

Tow heads and tow tails can turn up in 6 ways

i.e. THHT, HTHT, HHTT, HTTH, THTH, and TTHH ways

Hence the required probability $= \dfrac{6}{16}$

$$= \frac{3}{8}.$$

5. One of two events must happen: given that the chance of the one is two third that of the other find the odds in favor of the other

Let x be the probability of happening the second events then the probability of happening the first is $\dfrac{2}{3}x$.

Since one of the events must happen

$$\therefore \frac{2}{3}x + x = 1.$$

Or $\dfrac{5x}{3} = 1$

Or $x = \dfrac{3}{5}$

The probability of happening the second event is

$$x = \dfrac{3}{5} \; (\because 5 - 3 = 2$$

$$= \dfrac{3}{2+3}$$

Hence the odds in favor of the second event are 2 to 3.

6. **What is the chance of throwing an ace in the first only of two successive throws with an ordinary dice?**

There are 6 possible ways in which dice can fall, and of there one is favorable to the event that there is an ace in the first throw.

The chances of throwing an ace in the first show $= \dfrac{1}{6}$.

Similarly the chances of not throwing ace in the second throw $= 1 - \dfrac{1}{6}$

$$= \dfrac{5}{6}.$$

And since these events are independent of each other, therefore the required chances $= \dfrac{1}{6} \times \dfrac{5}{6}$

$$= \dfrac{5}{36}$$

7. **Three cards are drawn at random from an ordinary pack; find the chances that they will consist of a knave, a queen and a king**

The possible numbers cases in which three cards can be drawn from a pack of 52 cards are $^{52}C_3$.

The knave, queen and king can each be drawn in 4 ways each

Hence the favorable number of cases to the event required are $4 \times 4 \times 4 = 64$.

The required chances $= \dfrac{64}{^{52}C_3}$

$$= \dfrac{64}{\dfrac{52!}{(52-3)!\,3!}}$$

$$= \dfrac{64 \times 49! \times 3!}{52 \times 51 \times 50 \times 49!}$$

$$= \dfrac{\cancel{64}^{16} \times \cancel{6}^{2}}{\cancel{52}_{13} \times \cancel{51}_{17} \times \cancel{50}_{25}}$$

$$= \dfrac{16}{13 \times 17 \times 25}$$

$$= \dfrac{16}{5525}.$$

8. **The odds against a certain event are 5 to 2 and the odds in favor of the another event independent of the former are 6 to 5, find the chance that one at least of the events will happen.**

Given the odds against a first event are 5 to 2.

i.e. The chances that the first fails $= \dfrac{5}{7}$.

Again, the odds in favor of second event are 6 to 5

The chances that the second fails $= \dfrac{5}{11}$.

The chances that both fail $=\dfrac{5}{7} \times \dfrac{5}{11}$

$=\dfrac{25}{77}$

The two events are independent.

The chances that both do not fail, i.e.

Once at least of the events happen $=1-\dfrac{25}{77}$

$=\dfrac{77-25}{77}$

$=\dfrac{52}{77}$

9. **The odds against A, solving a certain problem are 4 to 3 and the odds in favor of B solving the same problem are 7 to 5, what is the chance that the problem will be solved if they both try?**

A' s chance of failure $=\dfrac{4}{7}$

And B' s chances of failure $=\dfrac{5}{12}$

The chances that both fail $=\dfrac{\cancel{4}^{1}}{7} \times \dfrac{5}{\cancel{12}_{3}}$ or $\dfrac{5}{21}$ for the two events are

independent

Therefore, the chances that both do not fail

i.e. The problem will be solved if they both try

$=1-\dfrac{5}{21}=\dfrac{21-5}{21}$

$=\dfrac{16}{21}$

10. What is the chance of drawing a rupee from a purse, one compartment of which contains 3 nP's and 2 rupees and the other 2 rupees and 1 nP?

As there are two compartments in the purse, either of the two may be selected,

Hence the chances of selecting the first compartment is $\dfrac{1}{2}$

Now in the first compartment there are in all 5 coins out of which 2 are rupees, then the chances of drawing a rupee is $\dfrac{2}{5}$. Therefore, the chances of a rupee from the first compartment are $\dfrac{1}{2} \times \dfrac{2}{5}$ or $\dfrac{1}{5}$; for the two events are dependent.

Similarly the chances of rupee from the other compartment is $\dfrac{1}{2} \times \dfrac{2}{3}$ or $\dfrac{1}{3}$

And since the two cases are mutually exclusive, the required chances

$$= \dfrac{1}{5} + \dfrac{1}{3}$$

$$= \dfrac{3+5}{15}$$

$$= \dfrac{8}{15}$$

11. There are four balls in a bag, but it is not known of what colors they are, one ball in drawn and found to be white, find the chance that all the balls are white.

If we consider that all numbers of white balls are in priori equally likely, we shall have 4 hypothesis.

1. All the balls in the bag may be white whose priori probability is denoted by P.

$$\Rightarrow P_1 = \frac{1}{4}.$$

2. Three of the may be white

Then whose probability is denoted by P_2

$$\Rightarrow P_2 = \frac{1}{4}$$

3. Two of them may be white

Whose probability is denoted by P_3

$$\Rightarrow P_3 = \frac{1}{4}$$

4. Only one of them may be white

Then whose priori probability is P_4

$$P_4 = \frac{1}{4}.$$

If all balls in the bag are white.

Then the chance drawing a white ball is $\frac{4}{4} = 1$.

If three of them are white, the chance of drawing a white balls is $\frac{3}{4}$.

If two balls are white, the chance of drawing a white ball is $\frac{2}{4} = \frac{1}{2}$.

If only one ball is white, the chance of drawing a white ball is $\frac{1}{4}$.

If only one ball is white, the chance of drawing a white ball is $\frac{1}{4}$.

Let these probabilities be denoted by p_1, p_2, p_3, p_4 respectively

Thus $p_1 = 1, \; p_2 = \frac{3}{4}, \; p_3 = \frac{1}{2}, \; p_4 = \frac{1}{4}.$

$$\therefore \; p_1 P_1 = 1 \times \frac{1}{4} = \frac{1}{4}$$

$$p_2 P_2 = \frac{3}{4} \times \frac{1}{4} = \frac{3}{16}$$

$$p_3 P_3 = \frac{1}{2} \times \frac{1}{4} = \frac{1}{8}$$

$$p_4 P_4 = \frac{1}{4} \times \frac{1}{4} = \frac{1}{16}$$

Hence the required chance $= \dfrac{p_1 P_1}{\sum pP} = \dfrac{\dfrac{1}{4}}{\dfrac{1}{4} + \dfrac{3}{16} + \dfrac{1}{8} + \dfrac{1}{16}}$

$$= \frac{\dfrac{1}{4}}{\dfrac{4 + 3 + 2 + 1}{16}}$$

$$= \frac{4}{10}$$

$$= \frac{2}{5}$$

12. **In a bag there are six balls of unknown colors, three balls are drawn and found to be black; find the chance that no black ball is left in the bag.**

We may consider here four hypothesis, all being equally likely.

(i) All the 6 balls in the bag may be black.

(ii) Five of them may be black

(iii) Four of them may be black.

(iv) only three balls may be black. Let the priori probabilities of there four hypothesis be denoted by P_1, P_2, P_3, P_4 respectively

Then $P_1 = P_2 = P_3 = P_4 = \dfrac{1}{4}$

Let P_1, P_2, P_3, P_4 denote the respecting probabilities that, when these hypothesis exist, the event of drawing 3 black balls follows;

Thus $p_1 = 1,\ p_2 = \dfrac{{}^5C_3}{{}^6C_3} = \dfrac{\dfrac{5!}{(5-3)!\,3!}}{\dfrac{6!}{(6-3)!\,3!}}$

$\qquad = \dfrac{5 \times 4}{2!} \times \dfrac{3!}{6 \times 5 \times 4}$

$\qquad = \dfrac{1}{2}$

$p_3 = \dfrac{{}^4C_3}{{}^6C_3} = \dfrac{\dfrac{4!}{(4-3)!\,3!}}{\dfrac{6!}{(6-3)!\,3!}} = \dfrac{\dfrac{4 \times 3!}{\cancel{3!}}}{\dfrac{6 \times 5 \times 4}{\cancel{3!}}}$

$\qquad = \dfrac{1}{5}$

$p_4 = \dfrac{1}{{}^6C_3} = \dfrac{\dfrac{1}{6!}}{(6-3)!\,3!} = \dfrac{\cancel{3!}\ \cancel{3!}}{\cancel{6} \times 5 \times 4 \times \cancel{3!}}$

$\qquad = \dfrac{1}{20}$

Because 3 balls out of 6 may be drawn in ${}^6C_3$ ways and 3 out of 5 in ${}^5C_3$, 3 out of 4 in ${}^4C_3$ and there is only one way of drawing 3 black balls from 3 black balls.

$\therefore\ p_1 P_1 = \dfrac{1}{4},\ \ p_2 P_2 = \dfrac{1}{8},\ \ p_3 P_3 = \dfrac{1}{20},\ \ p_4 P_4 = \dfrac{1}{80}$

And hence $\sum pP = \dfrac{35}{80}$

Therefore, the required chance $= \dfrac{p_4 P_4}{\sum pP}$

$$= \dfrac{\dfrac{1}{80}}{\dfrac{35}{80}} = \dfrac{1}{35}$$

13. A letter is known to have come from either London or Clifton, on the post mar only the two consecutive letters ON are legible, what is the chance that is come from London?

It is equally likely that letter may have come from Clifton on London. Let the priori probabilities of the event that the letter came from Clifton or London be denoted by P_1 and P_2 respectively.

Then $P_1 = P_2 = \dfrac{1}{2}$

Now, if the letter came from Clifton, there are 6 possible pairs of consecutive letters of which ON is one. Therefore the chance that this was the legible couple from the work Clifton is $\dfrac{1}{6}$.

Of the letter came from London, there are 5 possible pairs of consecutive letters of which two are ON. Therefore the chances that this was the legible couple from the London is $\dfrac{2}{5}$.

Let the probabilities of the event that the legible couple is from Clifton or London be denoted by P_1 and P_2 respectively.

Then $p_1 = \dfrac{1}{6}, \ p_2 = \dfrac{2}{5}$

$$\therefore p_1 P_1 = \dfrac{1}{6} \times \dfrac{1}{2} = \dfrac{1}{12}, \ p_2 P_2 - \dfrac{2}{5} \times \dfrac{1}{2} = \dfrac{1}{5}$$

Therefore the posterior chance that the letter was from London

$$= \frac{p_2 P_2}{\sum pP} = \frac{\dfrac{1}{5}}{\dfrac{1}{2}+\dfrac{1}{5}} = \frac{2}{7}$$

14. Before a race the chances of three runners A, B, C were estimated to be proportional to 5, 3, 2 but during the race A meets with an accident which reduces its chance to one third, what are now the respective chances of B and C?

The chances of A, B, C before the race starts are $\dfrac{5}{10}, \dfrac{3}{10}, \dfrac{2}{10}$ respectively. A can lose in two ways. Either by the winning of B or C. As A's chance of winning is $\dfrac{1}{2}$, therefore A's priori chance of losing is $\dfrac{1}{2}$.

But after the accident his chance of winning is $\dfrac{1}{3}$, and hence his chance of losing becomes $\dfrac{2}{8}$; that is his chance of losing is increased in the ratio of 4 to 3.

Therefore also B's and C's chances of winning are increased in the same ratio

Thus B's chance of winning $= \dfrac{3}{10} \times \dfrac{4}{3} = \dfrac{4}{10} = \dfrac{2}{5}$

And C's chance of winning $= \dfrac{2}{10} \times \dfrac{4}{3} = \dfrac{4}{15}$

15. A purse contains n coins of unknown value, a coin drawn at random is found to be a rupee, what is the chance that it is the only rupee in the purse.

If we consider that all numbers of rupees are a priori equally likely. We shall have n hypothesis; for (i) all the n coins may be rupees, $(i)^2 n - 1$

of them may be rupees $(i)^2 n - 2$ of them may be rupees... $(i)^{n-1}$ two of them may be rupees. $(i)^n$ Only one of them may be rupees. Let the priori probabilities of these hypothesis be denoted by $P_1, P_2, P_3, \cdots P_n$.

Then $P_1 = P_2 = P_3 = \cdots = P_n = \dfrac{1}{n}$.

Let $p_1, p_2, p_3, p_4 \cdots p_n$ denote the respective probabilities that the event of drawing a rupee will occur on the supposition that the rupee came from (i) or $(i)^2$, or, $\cdots$ or $(i)^n$

Thus $p_1 = \dfrac{n}{n}, \ p_2 = \dfrac{n-1}{n}, \ p_3 = \dfrac{n-2}{n}, \cdots$

$$P_{n-1} = \frac{2}{n}, \ p_n = \frac{1}{n}.$$

$$\therefore \ p_1 P_1 = \frac{n}{n} \times \frac{1}{n}, \ p_2 P_2 = \frac{n-1}{n} \cdot \frac{1}{n} \cdots$$

$$p_{n-1} P_{n-1} = \frac{2}{n} \cdot \frac{1}{n} \ \ p_n P_n = \frac{1}{n} \cdot \frac{1}{n}$$

$$\sum pP = \frac{1}{n} \left\{ \frac{n}{n} + \frac{n-1}{n} + \frac{n-2}{n} + \cdots + \frac{2}{n} + \frac{1}{n} \right\}$$

$$= \frac{1}{n} \cdot \frac{1}{n} \{ n + n - 1 + n - 2 + \cdots + 2 + 1 \}$$

$$= \frac{1}{n^2} \cdot \sum n$$

$$= \frac{1}{n^2} \frac{n(n+1)}{2}$$

$$= \frac{(n+1)}{2n}$$

Hence the required chance $= \dfrac{p_n P_n}{\sum pP} = \dfrac{\dfrac{1}{n^2}}{\dfrac{n+1}{2n}}$

$= \dfrac{2n}{n^2(n+1)}$

$= \dfrac{2}{n(n+1)}$

16. What are the odds in favor of throwing at least 7 in a single throw with two dice?

A throw amounting to 7 can be made of (1, 6) (2, 5) and (3, 4) each of which arrangements can occur in 2 ways; 8 can be made of (2, 6) (3, 5) and (4, 4) where each of the first two arrangements can occur in two ways and the last one in 1 way. 9 can be made of 3, 6 and 4, 5 each of which arrangements can occur ways. 10 can be made up of 4, 6 and 5, 5 where the first can occur in two ways and the last in one way; 11 can be made up of 5, 6 which can occur in two ways; and a throw amounting to 12 must be made of 6, 6 this can occur 1 way only.

Therefore the number of favorable cares to the event the sum on two faces amounts to at least 7 is $6+5+4+3+2+1$, or 21.

And the total number of cases is 6×6, or 36 are there are 6 possible ways of throwing a dice and the throwing of two dice being an independent event.

Thus the chance of throwing at least $7 = \dfrac{21}{36} = \dfrac{7}{12} = \dfrac{7}{7+5}$

Hence the odds in favor of throwing at least 7 in a single throw with two dice are 7 to 5.

17. In a purse there are 5 rupees and 4.50 np coins, if they are drawn out one by one, what is the chance that they came out rupees and 50 np coins alternatively, beginning with a rupee?

There are 9 coins in all in the purse out of which 5 are rupee and 4,50np coins.

Hence of chance of drawing a rupee $=\dfrac{5}{9}$; and the chance of drawing

a 50np coin when a rupee has already been drawn in $\dfrac{4}{8}$. The chance of

drawing a rupee when 850np coins and a 50np coin have already been

drawn is $\dfrac{4}{7}$ rupee. Similarly the chance of drawing 50np coin when two

rupees and one 50np coin have already been drawn is $\dfrac{3}{6}$; and now the

chance of drawing a rupee is $\dfrac{3}{5}$ and after it the chance of drawing a

50np coin is $\dfrac{2}{4}$.

The chance of drawing a rupee when three rupee and three 50np coin

have been drawn is $\dfrac{2}{3}$ and then the chance of drawing a 50np coins is

$\dfrac{1}{2}$ and then there is a certainly of drawing a rupee. Since all there

events are dependent

Hence the required chance

$$= \frac{\cancel{5}}{\cancel{9}_{3}} \times \frac{\cancel{4}}{\cancel{8}_{4}} \times \frac{\cancel{4}^{2}}{7} \times \frac{\cancel{3}}{\cancel{6}} \times \frac{\cancel{3}}{\cancel{5}} \times \frac{\cancel{2}}{\cancel{4}} \times \frac{1}{\cancel{2}} \times 1$$

$$= \frac{1}{126}$$

18. If on an average 9 ships out of 10 returns safe to port, what is the chance that out of 5 ships expected at least 3 will arrive?

The chance of the event that a ship returns safe to port is $\dfrac{9}{10}$. And

hence the chance that it does not return safe to the part is

$$1 - \frac{9}{10} \text{ or } \frac{10-9}{10} = \frac{1}{10}.$$

Three ships at least will return safe to the port in three mutually exclusive ways; if all the five are safe. Or 4 are safe, or 3 are safe, the chances of which are

$$\left(\frac{9}{10}\right)^5, {}^5C_4\left(\frac{9}{10}\right)^4\left(\frac{1}{10}\right), {}^5C_3\left(\frac{9}{10}\right)^3\left(\frac{1}{10}\right)^2 \text{ respectively.}$$

For all the five may arrive safe in one way, four will arrive safe when one does not arrive safe and others are safe and this can occur in ${}^5C_4$ different ways and similarly 3 will be safe when two of them are unsafe and this can occur in ${}^5C_3$ ways.

Therefore the required chance

$$= \left(\frac{9}{10}\right)^5 + {}^5C_4\left(\frac{9}{10}\right)^4\frac{1}{10} + {}^5C_3\left(\frac{9}{10}\right)^3\left(\frac{1}{10}\right)^2$$

$$= \frac{9^5}{10^5} + \frac{5!}{(5-4)!\,4!}\left(\frac{9}{10}\right)^4\frac{1}{10} + \frac{5!}{(5-3)!\,3!}\left(\frac{9}{10}\right)^3\left(\frac{1}{10}\right)^2$$

$$= \frac{9^5}{10^5} + 5\left(\frac{9}{10}\right)^4\frac{1}{10} + \frac{5\times4}{2}\left(\frac{9}{10}\right)^3\left(\frac{1}{10}\right)^2$$

$$= \frac{9^3}{10^5}\left\{9^2 + 5\times9 + 10\right\}$$

$$= \frac{729}{10^5}(81 + 45 + 10)$$

$$= \frac{729\times136}{10^5} = \frac{12393}{12500}$$

19. In a lottery all the tickets are blanks but one, each person draws a tickets and retains and retains it, show that each person has on equal chance of drawing the prize.

Let there be n tickets in all in the lottery, one of which is a prize and the others are blank.

Thus the chance of drawing a ticket containing prize $= \dfrac{1}{n}$, and hence the chance of drawing a blank ticket is $\dfrac{n-1}{n}$. If the first person fails to drawn the prize, there are now $n-1$ tickets one of which is prize, hence the chance of drawing the prize by second person is

$$\frac{n-1}{n} \cdot \frac{1}{n-1} = \frac{1}{n}$$

If the first two persons fail to draw the prize, the third person will draw from $n-2$ tickets, hence his chance of drawing the prize

$$= \frac{n-1}{n} \cdot \frac{n-2}{n-1} \cdot \frac{1}{n-2} \text{ or } \frac{1}{n}; \text{ and so on. Thus each chance is } \frac{1}{n} \text{ and}$$

hence each person has an equal chance of drawing the prize.

20. One bag contains 5 white and 3 red balls, and a second bag contains 4 white and 5 red balls. From one of them chosen at random, two balls are drawn; find the chance that they are of different colors.

Out of two bags any one may be chosen; the chance that the first bag is chosen is $\dfrac{1}{2}; \dfrac{5 \times 3}{^8C_2}$ or $\dfrac{15}{\dfrac{8!}{(8-2)!\,2!}} = \dfrac{15}{\dfrac{8 \times 7}{2}} = \dfrac{15}{28}.$

For two balls out of 8 may be drawn in $^8C_2$ ways and one white ball out of 5 may be drawn in 5 ways and similarly one red ball out of 3 may be drawn in 3 ways and these two when combined together give 5×3 different ways of drawing a white and a red ball.

Again the chance that the second bag is chosen is $\dfrac{1}{2}.$

And similarly the chance of choosing one white ball and one red from this bag is $\dfrac{4\times5}{^9C_2}$ or $\dfrac{20}{\dfrac{9!}{(9-2)!\,2!}}$

$$= \frac{20}{\dfrac{9\times8}{2}}$$

$$= \frac{20}{4\times9}$$

$$= \frac{20}{36}.$$

Since these two events are mutually exclusive therefore the required

chance $= \dfrac{1}{2}\times\dfrac{15}{28} + \dfrac{1}{\cancel{2}}\times\dfrac{\cancel{20}^{\,5}}{\cancel{36}_{18}}$

$$= \frac{15}{56} + \frac{5}{18}$$

$$= \frac{275}{504}$$

18 Determinants

1.   Consider the two homogeneous linear equations.

(a) $a_1 x + b_1 y = 0$

(b) $a_2 x + b_2 y = 0$

Multiplying the first equation (a) by b_2, the second equation (b) by b_1, subtracting and dividing by x, we obtain $a_1 b_2 - a_2 b_1 = 0$

This can be written in a compact form.

$$\begin{vmatrix} a_1 & b_1 \\ a_2 & b_2 \end{vmatrix} = 0$$

And the expression on the left is called a determinant.

A quick manipulation leads us to the following:

$$\begin{vmatrix} a_1 & b_1 \\ a_2 & b_2 \end{vmatrix} = a_1 b_2 - a_2 b_1 = \begin{vmatrix} a_1 & a_2 \\ b_1 & b_2 \end{vmatrix}$$

Observation 1: The value of the determinant is not altered by changing the rows into columns, and the columns into rows.

Again,

$$\begin{vmatrix} a_1 & b_1 \\ a_2 & b_2 \end{vmatrix} = - \begin{vmatrix} b_1 & a_1 \\ b_2 & a_2 \end{vmatrix} ; \begin{vmatrix} a_1 & b_1 \\ a_2 & b_2 \end{vmatrix} = - \begin{vmatrix} a_2 & b_2 \\ a_1 & b_1 \end{vmatrix}$$

Observation 2: If we interchange two rows or two columns of the determinant, we obtain a determinant which differs from it only in the sign.

2.   Consider the homogeneous linear equations:

$$a_1 x + b_1 y + c_1 z = 0$$
$$a_2 x + b_2 y + c_2 z = 0$$
$$a_3 x + b_3 y + c_3 z = 0$$

It expands to:

$$a_1 \begin{vmatrix} b_2 & c_2 \\ b_3 & c_3 \end{vmatrix} + b_1 \begin{vmatrix} c_2 & a_2 \\ c_3 & a_3 \end{vmatrix} + c_1 \begin{vmatrix} a_2 & b_2 \\ a_3 & b_3 \end{vmatrix} = 0$$

$$a_1(b_2 c_3 - b_3 c_2) + b_1(c_2 a_3 - c_3 a_2) + c_1(a_2 b_3 - a_3 b_2) = 0$$

By re-arranging terms the expanded form of the above determinant may be written

$$a_1(b_2 c_3 - b_3 c_2) + a_2(b_3 c_1 - b_1 c_3) + a_3(b_1 c_2 - b_2 c_1) = 0$$

Or

$$a_1 \begin{vmatrix} b_2 & b_3 \\ c_2 & c_3 \end{vmatrix} + a_2 \begin{vmatrix} b_3 & b_1 \\ c_3 & c_1 \end{vmatrix} + a_3 \begin{vmatrix} b_1 & b_2 \\ c_1 & c_2 \end{vmatrix} = 0 \qquad (1)$$

Hence:

$$\begin{vmatrix} a_1 & b_1 & c_1 \\ a_2 & b_2 & c_2 \\ a_3 & b_3 & c_3 \end{vmatrix} = \begin{vmatrix} a_1 & a_2 & a_3 \\ b_1 & b_2 & b_3 \\ c_1 & c_2 & c_3 \end{vmatrix}$$

Observation 3: The value of the determinant is not altered by changing the rows into columns, and the columns into rows.

$$\begin{vmatrix} a_1 & b_1 & c_1 \\ a_2 & b_2 & c_2 \\ a_3 & b_3 & c_3 \end{vmatrix} = a_1 \begin{vmatrix} b_2 & c_2 \\ b_3 & c_3 \end{vmatrix} - b_1 \begin{vmatrix} a_2 & c_2 \\ a_3 & c_3 \end{vmatrix} + c_1 \begin{vmatrix} a_2 & b_2 \\ a_3 & b_3 \end{vmatrix} \qquad (2)$$

From equation (1),we see that that coefficient of any one of the constituents a_1, a_2, a_3 is that determinant of the second order which is obtained by omitting the row and column in which it occurs. These determinants are called the Minors of the original determinants.

The left hand side of equation (1) may be written

$$a_1 A_1 - a_2 A_2 + a_3 A_3$$

Where A_1, A_2, A_3 are the minors of a_1, a_2, a_3 respectively

Again from equation (2), the determinant is equal to

$$a_1 A_1 - b_1 B_1 + c_1 C_1$$

Where A_1, B_1, C_1 are the minors of a_1, b_1, c_1 respectively.

3. The determinant is

$$\begin{vmatrix} a_1 & b_1 \\ a_2 & b_2 \end{vmatrix} = a_1(b_2 c_3 - b_3 c_2) + b_1(c_2 a_3 - c_3 b_2) + c_1(a_2 b_3 - a_3 b_2)$$

$$= -b_1(a_2 c_3) - a_1(c_2 b_3 - c_3 b_2) - c_1(b_2 a_3 - b_3 a_2)$$

Hence

$$\begin{vmatrix} a_1 & b_1 & c_1 \\ a_2 & b_2 & c_2 \\ a_3 & b_3 & c_3 \end{vmatrix} = - \begin{vmatrix} b_1 & a_1 & c_1 \\ b_2 & a_2 & c_2 \\ b_3 & a_3 & c_3 \end{vmatrix}$$

Observation 4: If two adjacent columns or rows of the determinant are interchanged, the sign of the determinant is changed, but its value remains un altered.

Observation 5: If two rows or two columns of the determinant are identical, the determinant vanishes.

Let:

$$D = \begin{vmatrix} a_1 & a_2 & a_3 \\ a_1 & a_2 & a_3 \\ b_1 & b_2 & b_3 \end{vmatrix}$$

$$-D = \begin{vmatrix} a_1 & b_1 & c_1 \\ a_2 & b_2 & c_2 \\ a_3 & b_3 & c_3 \end{vmatrix} \text{ (Interchanging rows 1 and 2, same result)}$$

$$D = -D \Rightarrow D = 0$$

Observation 6: If each constituent on any row or in any column is multiplied by the same factor, then the determinant is multiplied by that factor.

$$\begin{vmatrix} ma_1 & b_1 & c_1 \\ ma_2 & b_2 & c_2 \\ ma_3 & b_3 & c_3 \end{vmatrix} = ma_1 A_1 - ma_2 A_2 + ma_3 A_3$$

$$= m(a_1 A_1 - a_2 A_2 + a_3 A_3)$$

which proves the proposition.

Observation 7: If each constituent in any row, or column, consists of two terms, then the determinant can be expressed as the sum of two other determinants

Thus we have:

$$\begin{vmatrix} a_1 + \alpha_1 & b_1 & c_1 \\ a_2 + \alpha_2 & b_2 & c_2 \\ a_3 + \alpha_3 & b_3 & c_3 \end{vmatrix} = \begin{vmatrix} a_1 & b_1 & c_1 \\ a_2 & b_2 & c_2 \\ a_3 & b_3 & c_3 \end{vmatrix} + \begin{vmatrix} \alpha_1 & b_1 & c_1 \\ \alpha_2 & b_2 & c_2 \\ \alpha_3 & b_3 & c_3 \end{vmatrix}$$

For the expression on the left:

$$= (a_1 + \alpha_1) A_1 - (a_2 + \alpha_2) A_2 + (a_3 + \alpha_3) A_3$$

$$= (a_1A_1 - a_2A_2 + a_3A_3) + (\alpha_1A_1 - \alpha_2A_2 + \alpha_3A_3)$$
$$= \text{RHS}$$

Hence the proposition.

Similarly,

$$\begin{vmatrix} a_1 + \alpha_1 & b_1 + \beta_1 & c_1 \\ a_2 + \alpha_2 & b_2 + \beta_2 & c_2 \\ a_3 + \alpha_3 & b_3 + \beta_3 & c_3 \end{vmatrix} = \begin{vmatrix} a_1 & b_1 & c_1 \\ a_2 & b_2 & c_2 \\ a_3 & b_3 & c_3 \end{vmatrix} + \begin{vmatrix} \alpha_1 & b_1 & c_1 \\ \alpha_2 & b_2 & c_2 \\ \alpha_3 & b_3 & c_3 \end{vmatrix}$$

$$+ \begin{vmatrix} a_1 & \beta_1 & c_1 \\ a_2 & \beta_2 & c_2 \\ a_3 & \beta_3 & c_3 \end{vmatrix} + \begin{vmatrix} \alpha_1 & \beta_1 & c_1 \\ \alpha_2 & \beta_2 & c_2 \\ \alpha_3 & \beta_3 & c_3 \end{vmatrix}$$

Observation 8: Take the constituents of the row or column to be replaced, and increase or diminish them by any equimultiples of the corresponding constituents of one or more of the other rows and columns.

4. Now we will move on to the topic of determining the product of the determinants.

Let us start by investigating the value of

$$\begin{vmatrix} a_1\alpha_1 + b_1\beta_1 + c_1\gamma_1 & a_1\alpha_2 + b_1\beta_2 + c_1\gamma_2 & a_1\alpha_3 + b_1\beta_3 + c_1\gamma_3 \\ a_2\alpha_1 + b_2\beta_1 + c_2\gamma_1 & a_2\alpha_2 + b_2\beta_2 + c_2\gamma_2 & a_2\alpha_3 + b_2\beta_3 + c_2\gamma_3 \\ a_3\alpha_1 + b_2\beta_1 + c_2\gamma_1 & a_3\alpha_2 + b_3\beta_2 + c_3\gamma_3 & a_3\alpha_3 + b_3\beta_3 + c_3\gamma_3 \end{vmatrix}$$

We know that the above determinant can be expressed as the sum of 27 determinants of which it will be sufficient to give the following specimens:

$$\begin{vmatrix} a_1\alpha_1 & a_1\alpha_2 & a_1\alpha_3 \\ a_2\alpha_1 & a_2\alpha_2 & a_2\alpha_3 \\ a_3\alpha_1 & a_3\alpha_2 & a_3\alpha_3 \end{vmatrix}, \begin{vmatrix} a_1\alpha_1 & b_1\beta_2 & c_1\gamma_3 \\ a_2\alpha_1 & b_2\beta_2 & c_2\gamma_3 \\ a_3\alpha_1 & b_3\beta_2 & c_3\gamma_3 \end{vmatrix}, \begin{vmatrix} a_1\alpha_1 & c_1\gamma_2 & b_1\beta_3 \\ a_2\alpha_1 & c_2\gamma_2 & b_2\beta_3 \\ a_3\alpha_1 & b_3\gamma_2 & b_3\beta_3 \end{vmatrix}$$

These are respectively equal to:

$$\alpha_1\alpha_2\alpha_3 \begin{vmatrix} a_1 & a_1 & a_1 \\ a_2 & a_2 & a_2 \\ a_3 & a_3 & a_3 \end{vmatrix}, \alpha_1\beta_2\gamma_3 \begin{vmatrix} a_1 & b_1 & c_1 \\ a_2 & b_2 & c_2 \\ a_3 & b_3 & c_3 \end{vmatrix}, \alpha_1\beta_3\gamma_2 \begin{vmatrix} a_1 & c_1 & b_1 \\ a_2 & c_2 & b_2 \\ a_3 & c_3 & b_3 \end{vmatrix}$$

The first of which vanishes. Similarly it will be found that 21 out of the 27 determinants vanish. The six determinants that remain are equal to

$$\begin{pmatrix} \alpha_1\beta_2\gamma_3 - \alpha_1\beta_3\gamma_2 + \alpha_2\beta_3\gamma_1 \\ -\alpha_2\beta_1\gamma_3 + \alpha_3\beta_1\gamma_2 - \alpha_3\beta_2\gamma_1 \end{pmatrix} \times \begin{vmatrix} a_1 & b_1 & c_1 \\ a_2 & b_2 & c_2 \\ a_3 & b_3 & c_3 \end{vmatrix}$$

That is:

$$\begin{vmatrix} \alpha_1 & \beta_1 & \gamma_1 \\ \alpha_2 & \beta_2 & \gamma_2 \\ \alpha_3 & \beta_3 & \gamma_3 \end{vmatrix} \times \begin{vmatrix} a_1 & b_1 & c_1 \\ a_2 & b_2 & c_2 \\ a_3 & b_3 & c_3 \end{vmatrix}$$

Hence the given determinant can be expressed as the product of two other determinants.

Consider the two linear equations:

$$a_1 X_1 + b_1 X_2 = 0$$
$$a_2 X_2 + b_2 X_2 = 0 \tag{1}$$

Where:

$$X_1 = \alpha_1 x_1 + \alpha_2 x_2$$
$$X_2 = \beta_1 x_1 + \beta_2 x_2 \tag{2}$$

Substituting for X_1, X_2 in (1), we have,

$$(a_1\alpha_1 + b_1\beta_1)x_1 + (a_1\alpha_2 + b_1\beta_2)x_2 = 0$$
$$(a_2\alpha_1 + b_2\beta_1)x_1 + (a_2\alpha_2 + b_2\beta_2)x_2 = 0 \tag{3}$$

In order that equations (3) may simultaneously hold for values of x_1 and x_2 other than zero, we must have,

$$\begin{vmatrix} a_1\alpha_1 + b_1\beta_1 & a_1\alpha_2 + b_1\beta_2 \\ a_2\alpha_1 + b_2\beta_2 & a_2\alpha_2 + b_2\beta_2 \end{vmatrix} = 0 \tag{4}$$

But equations (3) will hold if equations (1) hold, and this will be the case either if:

$$\begin{vmatrix} a_1 & b_1 \\ a_2 & b_2 \end{vmatrix} = 0 \tag{5}$$

Or if $X_1 = 0$ and $X_2 = 0$ which last condition requires that

$$\begin{vmatrix} \alpha_1 & \alpha_2 \\ \beta_1 & \beta_2 \end{vmatrix} = 0 \tag{6}$$

Hence if equations (5) and (6) hold, equation (4) must also hold; and therefore the determinant in (4) must contain as factors the determinants in (5) and (6); and a consideration of the dimensions of the determinants shows that the remaining factor of (4) must be numerical. Hence

$$\begin{vmatrix} a_1 & b_1 \\ a_2 & b_2 \end{vmatrix} \times \begin{vmatrix} \alpha_1 & \beta_1 \\ \alpha_2 & \beta_2 \end{vmatrix} = \begin{vmatrix} a_1\alpha_1 + b_1\beta_1 & a_1\alpha_2 + b_1\beta_2 \\ a_2\alpha_1 + b_2\beta_2 & a_2\alpha_2 + b_2\beta_2 \end{vmatrix}$$

The numerical factor, by comparing the coefficients of $a_1 b_2 \alpha_1 \beta_2$ on the two sides of the equations, is seen to be unity.

Observation 9:

$$\begin{vmatrix} a_1 & b_1 \\ a_2 & b_2 \end{vmatrix}^2 = \begin{vmatrix} a_1^2 + b_1^2 & a_1 a_2 + b_1 b_2 \\ a_1 a_2 + b_1 b_2 & a_2^2 + b_2^2 \end{vmatrix}$$

5. Let us have equations:

$$a_1 x + b_1 y + c_1 z + d_1 = 0$$
$$a_2 x + b_2 y + c_2 z + d_2 = 0$$
$$a_3 x + b_3 y + c_3 z + d_3 = 0$$

Let:

$$D = \begin{vmatrix} a_1 & b_1 & c_1 \\ a_2 & b_2 & c_2 \\ a_3 & b_3 & c_3 \end{vmatrix}$$

The solution may be written as:

$$\frac{x}{\begin{vmatrix} d_1 & b_1 & c_1 \\ d_2 & b_2 & c_2 \\ d_3 & b_3 & c_3 \end{vmatrix}} = \frac{-y}{\begin{vmatrix} d_1 & a_1 & c_1 \\ d_2 & a_2 & c_2 \\ d_3 & a_3 & c_3 \end{vmatrix}} = \frac{z}{\begin{vmatrix} d_1 & a_1 & b_1 \\ d_2 & a_2 & b_2 \\ d_3 & a_3 & b_3 \end{vmatrix}} = \frac{-1}{\begin{vmatrix} a_1 & b_1 & c_1 \\ a_2 & b_2 & c_2 \\ a_3 & b_3 & c_3 \end{vmatrix}}$$

6. Suppose we have the system of four homogeneous linear equations:

$$a_1 x + b_1 y + c_1 z + d_1 u = 0$$
$$a_2 x + b_2 y + c_2 z + d_2 u = 0$$
$$a_3 x + b_3 y + c_3 z + d_3 u = 0$$
$$a_4 x + b_4 y + c_4 z + d_4 u = 0$$

From the last three of these, we have as in the preceding article,

$$\frac{x}{\begin{vmatrix} b_2 & c_2 & d_2 \\ b_3 & c_3 & d_3 \\ b_4 & c_4 & d_4 \end{vmatrix}} = \frac{-y}{\begin{vmatrix} a_2 & c_2 & d_2 \\ a_3 & c_3 & d_3 \\ a_4 & c_4 & d_4 \end{vmatrix}} = \frac{z}{\begin{vmatrix} a_2 & b_2 & d_2 \\ a_3 & b_3 & d_3 \\ a_4 & b_4 & d_4 \end{vmatrix}} = \frac{-u}{\begin{vmatrix} a_2 & b_2 & c_2 \\ a_3 & b_3 & c_3 \\ a_4 & b_4 & c_4 \end{vmatrix}}$$

7. More generally, if we have n homogeneous linear equations

$$a_1 x_1 + b_1 x_2 + c_1 x_3 + \ldots + k_1 x_n = 0$$
$$a_2 x_1 + b_2 x_2 + c_2 x_3 + \ldots + k_2 x_n = 0$$
$$a_3 x_1 + b_3 x_2 + c_3 x_3 + \ldots + k_3 x_n = 0$$

...

$$a_n x_1 + b_n x_2 + c_n x_3 + \ldots + k_n x_n = 0$$

Involving n unknown quantities $x_1, x_2, x_3, \ldots, x_n$ these quantities can be eliminated and the result expressed in the form

$$\begin{vmatrix} a_1 & b_1 & c_1 & \ldots & k_1 \\ a_2 & b_2 & c_2 & \ldots & k_2 \\ a_3 & b_3 & c_3 & \ldots & k_3 \\ \ldots & \ldots & \ldots & \ldots & \ldots \\ a_n & b_n & c_n & \ldots & k_n \end{vmatrix} = 0$$

The left hand number of this equation is a determinant which consists of n rows and n columns, and is called a determinant of the nth order.

18.1 Solved Problems

Calculate the values of the determinants:

1.
$$\begin{vmatrix} 1 & 1 & 1 \\ 35 & 37 & 34 \\ 23 & 26 & 25 \end{vmatrix}$$

$$\begin{vmatrix} 1 & 1 & 1 \\ 35 & 37 & 34 \\ 23 & 26 & 25 \end{vmatrix} = \begin{vmatrix} 1 & 1-1 & 1-1 \\ 35 & 37-35 & 34-35 \\ 23 & 26-23 & 25-23 \end{vmatrix}$$

$$= \begin{vmatrix} 1 & 0 & 0 \\ 35 & 2 & -1 \\ 23 & 3 & 0 \end{vmatrix} = \begin{vmatrix} 2 & -1 \\ 3 & 2 \end{vmatrix} = 4+3 = 7$$

2. $$\begin{vmatrix} 13 & 16 & 19 \\ 14 & 17 & 20 \\ 15 & 18 & 21 \end{vmatrix}$$

$$\begin{vmatrix} 13 & 16 & 19 \\ 14 & 17 & 20 \\ 15 & 18 & 21 \end{vmatrix} = \begin{vmatrix} 13 & 16-13 & 19-13 \\ 14 & 17-14 & 20-14 \\ 15 & 18-15 & 21-15 \end{vmatrix}$$

$$= \begin{vmatrix} 13 & 3 & 6 \\ 14 & 3 & 6 \\ 15 & 3 & 6 \end{vmatrix} = \begin{vmatrix} 13 & 3 & 6-2\times3 \\ 14 & 3 & 6-2\times3 \\ 15 & 3 & 6-2\times3 \end{vmatrix} = \begin{vmatrix} 13 & 3 & 0 \\ 14 & 3 & 0 \\ 15 & 3 & 0 \end{vmatrix}$$

$$= 0$$

3. $$\begin{vmatrix} 13 & 3 & 23 \\ 30 & 7 & 53 \\ 39 & 9 & 70 \end{vmatrix}$$

$$\begin{vmatrix} 13 & 3 & 23 \\ 30 & 7 & 53 \\ 39 & 9 & 70 \end{vmatrix} = \begin{vmatrix} 13-4\times3 & 3 & 23-7\times3 \\ 30-4\times7 & 7 & 53-7\times7 \\ 39-4\times9 & 9 & 70-7\times9 \end{vmatrix}$$

$$= \begin{vmatrix} 13 & 3 & 2 \\ 30 & 7 & 4 \\ 39 & 9 & 7 \end{vmatrix} = \begin{vmatrix} 1 & 0 & 0 \\ 2 & 1 & 0 \\ 3 & 0 & 1 \end{vmatrix} = 1$$

4.
$$\begin{vmatrix} 1 & 1 & 1 & 1 \\ 1 & 2 & 3 & 4 \\ 1 & 3 & 6 & 10 \\ 1 & 4 & 10 & 20 \end{vmatrix}$$

$$\begin{vmatrix} 1 & 1 & 1 & 1 \\ 1 & 2 & 3 & 4 \\ 1 & 3 & 6 & 10 \\ 1 & 4 & 10 & 20 \end{vmatrix} = \begin{vmatrix} 1 & 0 & 0 & 0 \\ 1 & 1 & 2 & 3 \\ 1 & 2 & 5 & 9 \\ 1 & 3 & 9 & 19 \end{vmatrix}$$

$$= \begin{vmatrix} 1 & 2 & 3 \\ 2 & 5 & 9 \\ 3 & 9 & 19 \end{vmatrix} = \begin{vmatrix} 1 & 0 & 0 \\ 2 & 1 & 3 \\ 3 & 3 & 9 \end{vmatrix} = \begin{vmatrix} 1 & 3 \\ 3 & 10 \end{vmatrix} = \begin{vmatrix} 1 & 0 \\ 3 & 1 \end{vmatrix}$$

$$= 1$$

5.
$$\begin{vmatrix} 7 & 13 & 10 & 6 \\ 5 & 9 & 7 & 4 \\ 8 & 12 & 11 & 7 \\ 4 & 10 & 6 & 3 \end{vmatrix}$$

$$\begin{vmatrix} 7 & 13 & 10 & 6 \\ 5 & 9 & 7 & 4 \\ 8 & 12 & 11 & 7 \\ 4 & 10 & 6 & 3 \end{vmatrix} = \begin{vmatrix} 1 & 1 & 4 & 6 \\ 1 & 1 & 3 & 4 \\ 1 & -2 & 4 & 7 \\ 1 & 4 & 3 & 3 \end{vmatrix} = \begin{vmatrix} 1 & 0 & 0 & 2 \\ 1 & 0 & -1 & 1 \\ 1 & -3 & 0 & 3 \\ 1 & 3 & -1 & 0 \end{vmatrix}$$

$$= \begin{vmatrix} 1 & 0 & 0 & 0 \\ 1 & 0 & -1 & -1 \\ 1 & -3 & 0 & 1 \\ 1 & 3 & -1 & 2 \end{vmatrix} = \begin{vmatrix} 0 & -1 & -1 \\ -3 & 0 & 1 \\ 3 & -1 & -2 \end{vmatrix} = \begin{vmatrix} 0 & 0 & -1 \\ -3 & -1 & 1 \\ 3 & +1 & 2 \end{vmatrix}$$

$$= -1 \begin{vmatrix} -3 & -1 \\ 3 & 1 \end{vmatrix} = 0$$

6.
$$\begin{vmatrix} a & 1 & 1 & 1 \\ 1 & a & 1 & 1 \\ 1 & 1 & a & 1 \\ 1 & 1 & 1 & a \end{vmatrix}$$

$$\begin{vmatrix} a & 1 & 1 & 1 \\ 1 & a & 1 & 1 \\ 1 & 1 & a & 1 \\ 1 & 1 & 1 & a \end{vmatrix} = \begin{vmatrix} a-1 & 0 & 0 & 1 \\ 0 & a-1 & 0 & 1 \\ 0 & 0 & a-1 & 1 \\ 1-a & 1-a & 1-a & a \end{vmatrix}$$

$$= (a-1)^3 \begin{vmatrix} 1 & 0 & 0 & 1 \\ 0 & 1 & 0 & 1 \\ 0 & 0 & 1 & 1 \\ -1 & -1 & -1 & a \end{vmatrix} = (a-1)^3 \begin{vmatrix} 1 & 0 & 0 & 0 \\ 0 & 1 & 0 & 1 \\ 0 & 0 & 1 & 1 \\ -1 & -1 & -1 & a+1 \end{vmatrix}$$

$$= (a-1)^3 \begin{vmatrix} 1 & 0 & 1 \\ 0 & 1 & 1 \\ -1 & -1 & a+1 \end{vmatrix} = (a-1)^3 \begin{vmatrix} 1 & 0 & 0 \\ 0 & 1 & 1 \\ -1 & -1 & a+2 \end{vmatrix}$$

$$= (a-1)^3 \begin{vmatrix} 1 & 1 \\ -1 & a+2 \end{vmatrix} = (a-1)^3 \begin{vmatrix} 1 & 0 \\ -1 & a+3 \end{vmatrix}$$

$$= (a-1)^3 (a+3)$$

19 Closing Thoughts

Mathematics is not a spectator sport. The patterns and underlying nuances are like a work of art. The more you apply yourself to the subject, the more you uncover and understand. Mathematics is a subject which requires practice. This is not something that you relax on your couch, casually browse through and hope to achieve mastery. This will require patience and application.

1. Neatness is conducive to accuracy. Refrain from the temptation to write down something quickly and scratching the same to make the necessary corrections.

2. One of the weakness we find in student while solving word problems is the usage of $=$ sign. This sign as a specific meaning in the world of mathematics. It cannot be used as a way to begin every new line of step in the problem solving process. Use appropriate mathematical signs and symbols. Never use them to mean something vague. $=$ Sign is never good space filler.

3. Spend a second or two to explain how you arrived at a certain step. Several books and references use a statement, such as ``it follows from the above statement". We have oftentimes wondered how the expression or equation below follows from the one above. A good explanation is an excellent demonstration of your understanding of the underlying principles.

4. When you are faced with several conclusions during a problem solving process, it is a good idea to number the statements or equations. In subsequent steps, you can refer to these conclusions by using the label or the assigned equation number.

5. The easiest of problems attracts the silliest of mistakes. If the problem is easy, motivate yourself to get it right. Do not let overconfidence or carelessness to take control of the situation

www.ingramcontent.com/pod-product-compliance
Lightning Source LLC
Chambersburg PA
CBHW071415170526
45165CB00001B/282